THE ALPACA HANDBOOK

The Complete Guide to Raising a Happy and Healthy Herd

REBECCA GILL
of Cotton Creek Farms

LYONS PRESS

ESSEX, CONNECTICUT

An imprint of The Globe Pequot Publishing Group, Inc.
64 South Main Street
Essex, CT 06426
www.globepequot.com

British Library Cataloguing in Publication Information available

Library of Congress Cataloging-in-Publication Data available
ISBN 978-1-4930-9227-7 (cloth: alk. paper)
ISBN 978-1-4930-9228-4 (electronic)

Printed in India

CONTENTS

INTRODUCTION

Raising Alpacas for Fun, Love, and Profit

What happens when a city girl working in technology moves to the country and starts an alpaca farm? Magic happens—and this magic could only be delivered by a herd of sweet, funny, and completely lovable alpacas. Alpacas are quirky animals that possess an amazing ability to deliver all the things that really matter in life. They're peaceful and entertaining, and they provide a whole new meaning to the phrase "work-life balance."

I've gone from speaking at business conferences to sticking my arm inside an alpaca to help with a difficult birth. My Egyptian cotton sheets are now used for alpaca births. At the end of each day, my feet are dirty and my hair smells of barn. I've packed up my business suits and replaced them with work boots and alpaca socks. I've said goodbye to "fancy," and I've opened myself up to a rich life full of barns, hay, and alpaca kisses.

In my life, I've had moments of extreme certainty that something was just right. It's an instinctual feeling that I sense at my core. Moving north and starting our alpaca farm were both decisions that needed no thought, discussion, or deliberation. In my heart, I knew it was the right thing for me and my family. This change brought color to a black-and-white life. I felt like I could breathe in fully and without reservation. And I don't think this change came from wide-open spaces and fresh country air.

Raising alpacas keeps me centered and grounded, which allows me to worry less and experience life more. It's also brought the amazing gifts of coffee in the pasture, sunsets on the front porch, and a sense of peace in my soul. And I am happy.

Now I want to bring this happiness to others. When asked to write a book about raising alpacas, I didn't hesitate. I want to help others find that "alpaca magic" I've discovered. I want the world to know that raising alpacas can help you feel loved, grounded, and complete. I want you to feel that same magic, happiness, and peace that alpacas have brought to my life.

Cotton Creek
FARMS
Alpaca Store • Interactive Tours
Sales • Breeding • Boarding
WWW.COTTONCREEKFARMS.COM

In My Life, All Roads Have Led to Alpacas

My life has not been an easy one. I grew up in a lot of chaos and faced a lot of loss. I longed for stability, love, and a sense of belonging. When I secured my first job, I felt as though I found some stability. When I married my husband, Jason, I knew I had found love. When I gave birth to my two children, I finally felt like I had a sense of belonging. And yet through all that discovery and growth, something was still missing.

My husband and I were successful professionals who lived in a fancy subdivision, our children attended a top-ranked school district, and we even owned a large vacation home on a beautiful lake. To everyone around us, it looked like we were living the perfect life. And yet, life didn't feel perfect or complete. My husband and I felt like we were suffocating. We simply never had a moment to pause and breathe. Worse yet, our son was miserable, and we were watching him slip into a downward

spiral. He hated the large school, busy schedule, and all the elements that came with living in the suburbs.

So, we paused. We took a family vote, and we decided we were going to sell everything we owned and move to northern Michigan. I grew up there, so I knew what awaited us. Small-town vibes, family-owned farms, lots of snow, and a slower pace of life. I've long said I was raised by a village, and that is because I grew up in a small community where people knew my story and went out of their way to make my life better. It was that environment I now wanted to offer to my son. He certainly didn't need the help I did as a child, but I knew the small-town community would help him thrive.

When we announced we were leaving metro Detroit to start a farm, virtually everyone around us thought something was wrong. To most people, the idea sounded completely absurd. To us, it was a purposeful decision to create "Life 2.0." We honestly didn't know what that life would entail, but we hoped it would offer a slower pace that included purpose.

It took us a year to find a twenty-seven-acre parcel that was located south of Traverse City. As soon as we found it, we submitted a cash offer and started to put our Life 2.0 plan into action. At this point we still had no idea what we were going to do once we finally moved into our country home.

Then one day we randomly stopped at an alpaca farm. This changed *everything*. I was absolutely captivated by these magical animals. The alpacas were just like the funny, quirky Muppets I grew up with in the 1970s, but they were alive, and they loved me. My husband did not feel the same, but he appeased me by discussing the possibility of starting an alpaca farm.

While I was at a work conference, Jason visited a few more alpaca farms and researched owning alpacas as a business. After a lot of discussion and a family debate, we purchased five female alpacas. They were nothing fancy, but they were the start of something much bigger. I rarely get excited about anything in life, and I've always been that way. This was different. I was *really* excited about buying and bringing home our alpacas. Let's just say this city girl had no idea what was about to transpire!

Five Alpacas Quickly Led to Seventy Alpacas

A few weeks after owning our first five alpacas, the local 4-H office asked if we could adopt two more alpacas that needed homes. We did and soon five became seven. When Faith and Stormy arrived, it was like looking at the child I once was. They were sweet alpacas who needed an opportunity to feel safe and loved. I knew if I could give them this, their true personalities would shine. Stormy fell in love with me immediately. She would jump and dance around when she'd see me walking to the barn, which quickly showed me the true personality of an alpaca. It took me many months to obtain the trust of Faith, but I did, and today she is happy and well adjusted. This only reinforced what I already knew: Alpacas are unique, quirky creatures who are smart, loving, and lots of entertainment. This made my love for them grow deeper and richer than I thought was possible.

And then I blinked, and our small farm quickly grew into seventy alpacas. Then I blinked again, and our small farm now had an alpaca store, we were running an active agritourism business, we were breeding and selling alpacas, we were guest speakers at alpaca events, and we were even manufacturing our own line of alpaca products.

There was a lot I didn't know back then, and there is even more I still don't know today. The city girl in me still conflicts with the farmer, but the farmer rises to the occasion.

Everyone who visits our farm asks why we decided to raise alpacas. And while I can't remember how we first started talking about them, I can tell you that in my core I knew an alpaca farm was the right decision. When our friends and family come to visit, they see the change in me, my husband, and our son. They see the happiness and peace we have, they see the happiness our farm brings to others, and they see why alpacas are the perfect animal and business for our family.

So that's my story. Now let's talk about how you can get started with your alpaca story.

CHAPTER 1

Is the Alpaca Lifestyle Right for You and Your Farm?

There is a romantic appeal to alpacas that allows them to find their way into Hallmark movies and Instagram videos. I've felt that romance and I can 100 percent confirm it exists. I love my herd and the lifestyle they've helped me create. And much to my surprise, that romance seems to continue even when I'm sick with a cold and I still need to go clean a barn, feed hay, or check on a pregnancy.

I can't count how many times I've walked out to the barn in my pajamas and flip-flops at nine o'clock at night to check on someone or something. I can be tired and ready for bed, yet something propels me out the door. I'm emotionally invested in

my alpaca herd, and I'll tend to them just as I would my human children. I remember being absolutely exhausted taking care of my sick children, only to have my energy rebound when my sick son smiles or gives me a hug. I feel the same on those late-night trips to the barn when everyone comes to greet me, tug on my clothes, or reach in for a kiss. I know they appreciate me, the care I provide, and the love I offer.

Make no mistake—alpacas and their love fuel me. Their quirky personalities entertain me. Their smart and sassy attitudes enrich everything in me. It's how the romantic allure of owning alpacas grabs you and holds you tight.

But as you consider raising alpacas, remember not just to look at the romance, but also to think about the lifestyle they offer and whether that lifestyle is right for you.

Alpacas Can Help You Create the Life You Want and Need

When alpacas were first imported into the United States, they were marketed as a "lifestyle business" offering endless financial opportunities for young families and retired couples. It was an easy value proposition to make since alpacas are one of the few livestock animals that are gentle, easy to care for, and offer a diverse revenue stream. For those who want to make alpacas an investment, you can earn a great living selling alpacas, producing alpaca products to sell, or engaging in agritourism activities like interactive tours.

Decades ago, the "I Love Alpacas" campaign used the phrase "the world's finest livestock investment" to market them to future owners. Yes, owning alpacas does offer a plethora of revenue opportunities and personal enjoyment, but the truth is, it isn't for everyone. These animals are unique in biology, personality, and care, which brings its own set of benefits and challenges. That means raising alpacas isn't right for every person, family, or farm, but it could be *perfect for you*.

When my husband and I made the decision to move to the country, we did so with purpose. We wanted to create a different lifestyle for our family, and we have kept that in mind as our journey with alpacas has evolved. The one constant we have kept in place is the lifestyle goal we set, and we continue to tweak our farm to make sure we stay true to those goals.

PHOTO PROVIDED BY LAUREN ALYSSA PHOTOGRAPHY

Thankfully, alpacas were an ideal fit for the lifestyle we wanted, and we have been immensely rewarded in raising them. No other livestock animal fit our needs like alpacas have.

My goal is to provide education to help future alpaca owners determine if raising alpacas is right for their situation and their goals. To do that, I'd like to start with a simple list of pros and cons.

Alpacas are a member of the camelid family and originate from South America. They were domesticated over six thousand years ago and arrived in the United States in 1984. Today, about 85 percent of the world's alpaca population still lives in the Peruvian highlands.

An alpaca can weigh between one hundred and two hundred pounds, live to be between fifteen and twenty years old, and produce about five to ten pounds of fiber per year. They grow fiber in a variety of colors and produce fur that is ultra-soft, sustainable, hypoallergenic, and extremely warm.

The Pros of Raising Alpacas

No matter how cute and fluffy alpacas are, you should weigh the pros and cons of raising them before you bring any home. Alpacas have a lot to offer, but they come with some nuances that need to be considered.

Alpacas are easy keepers, and all things considered, they are much easier to raise than other livestock. When we owned five alpacas, chores only took about fifteen minutes. My husband ended up spending a lot more time with them, but he wasn't doing chores. He was just hanging out with the herd and enjoying his time with them.

When we reached seventy alpacas, we realized those easy keepers were more work than our family could manage on its own, so we scaled back the herd size to reset and find greater balance. This allowed us to move back toward that lifestyle we wanted and the work-life balance we set out to establish.

Alpacas are gentle and peaceful animals, so they're a great option for interacting with children, the elderly, and those with disabilities. Since we opened our farm to the public, we've probably had over twenty thousand people come through our farm. We've never had to worry about anyone with special needs interacting with our herd because the alpacas can sense something is different and they adapt their behavior to match this need.

We've had many small children on our farm, and we've never had a safety issue because the alpacas can sense their size and vulnerability. I was once giving a tour, and a toddler crawled into our adult male pen when the parents were distracted. The mom panicked but I found myself smiling. I calmly walked over, opened the gate, scooped up the toddler, and handed him back to his parents. I knew the child was safe, because I was confident my boys would be curious but never dangerous. And they were just that. My herd views humans as friends, so aggression would never come into the equation.

Alpacas are smart and very trainable, which makes it easy to modify their behavior when needed. Our herd knows the difference between me, my husband, and our son. Just like human children, they know what they can get away with and they know this varies by human.

Our herd is on their best behavior around me because they know I expect it, and they want to please me. When my son is in the barn, not so much. They'll argue over feed and virtually do whatever they want, because they are smart enough to know he'll tolerate it.

I treat our herd like young children. I am consistent in my expectations and my reactions to poor behavior. Without realizing I was doing it, I trained them to notice my "mom look" so they know when their behavior crosses the line. They are fast to course correct, which makes my interactions pleasant.

Alpacas are sustainable and easy on the earth, so they won't damage lawns or pastures like cows, pigs, goats, or horses. Alpacas nibble on grass; they don't pull on it like other livestock. This keeps grass rooted and healthy. They also have soft, padded feet that don't damage the ground, soil, or grass. They leave the grass looking like they found it, just a wee bit shorter.

Alpacas don't require a lot of space. While they do like to occasionally run, it is short lived. Our alpacas will run out to the pasture when it is time to graze, but that running only lasts for about thirty seconds. For the most part, they need enough space to graze and get exercise.

Alpacas are cheaper to care for than the family dog. Since alpacas are efficient eaters, they have a much smaller intake of food than a traditional livestock animal. Overall hay costs are a fraction of what they would be for other livestock. And unlike your dog who needs annual vet visits, you only need to have your vet examine your alpaca if something is wrong. For many alpacas, nothing ever goes wrong, so vet bills are nonexistent.

Alpacas use communal poop piles, so cleanup is much faster than with other livestock. By default, alpacas will create specific areas where they pee and poop. One alpaca declares the spot and the others follow. This makes cleaning up fast, and also makes it easy to control fly production.

Alpaca fiber is known as the "fiber of the gods." Alpacas produce amazing fiber that can be used in clothing, home goods, toys, and lots of other items. It is as soft as cashmere, three times warmer than sheep's wool, and hypoallergenic. These qualities make it a highly desirable material for all types of use and for all types of people.

Raising alpacas provides multiple revenue streams. While not everyone will purchase alpacas for business purposes, revenue options are plentiful should this be of interest to you. Alpaca farmers can decide which revenue stream works for their lifestyle. From breeding and fiber products to agritourism and services such as renting out alpacas for weddings and photo shoots, there is a revenue option for every farm.

Owning alpacas can reduce income taxes. The US government classified alpacas as livestock back in 2008, which opened the door to tax breaks. As a small farm you can save money on taxes via property tax reductions, expensing vehicles, writing off utilities for business, and so on. You can write off a lot more, but I'll leave the rest of this discussion to you and your accountant.

The Cons of Raising Alpacas

I strongly believe the pros of owning alpacas outweigh the cons; however, I wouldn't do this discussion justice if I didn't present some cons to consider.

Alpacas are herd animals, so you should start with three of the same sex alpacas for them to be mentally and physically healthy. While this is the industry benchmark, we've also learned firsthand why this is the recommendation, as I'll talk more about in future chapters. We've helped people correctly size their herds after they tried to deviate from the recommended herd size and ran into trouble.

Males and females must be kept separate. The female alpaca does not have a traditional "cycle," so breeding can take place at any time. This makes keeping males and females together physically impossible.

With some livestock you can "geld" the male and this resolves the issue. This is not true for alpacas, and even gelded males cannot be housed with females. Gelding only takes away the ability to produce offspring; it does not take away the ability for the male to have intercourse. The female anatomy cannot withstand frequent breeding, which leads to the need to separate males and females.

Alpacas should not be housed with other livestock due to parasites and behavior concerns. Chickens, cats, and some livestock dogs (like Anatolian shepherds or Great Pyrenees) are suitable companions, but other livestock animals create an opportunity for issues. The alpaca immune system isn't designed to manage parasites from other animals, so these foreign parasites can quickly lead to illness and/or death.

Alpacas also are lovers and not fighters. They don't have the ability to defend themselves against a playful goat or a kicking horse.

Something to keep in mind is that this doesn't mean your farm cannot have other livestock. It simply means the other livestock should have their own areas for shelter and grazing.

Alpacas require proper infrastructure of shelter, fencing, and access to food and water. Requirements will vary based on where you live and how many alpacas you own. For example, in the state of Michigan, you'd need electricity run to your barn for heating water buckets in winter and operating fans in the heat of summer. In areas with consistently mild weather, electricity wouldn't be required.

Because alpacas are classified as livestock, they cannot live in a traditional backyard. They must reside in locations designated as agricultural. Something to keep in mind is there are different types of zoning that allow livestock, although special-use permits can sometimes be obtained for deviations. Checking with your local planning office will help you know what zone you live in and if there are other local requirements that need to be adhered to.

Not all livestock vets treat alpacas. While a herd raised solely for their fiber will rarely need medical care, breeding alpacas will, and you'll need an experienced alpaca

vet that is close by to assist in emergencies. An alternative to this would be access to a university that teaches alpaca care or an experienced mentor. There are many alpaca owners who live in remote areas without veterinary assistance, and the key to their success is having a plan for emergency care.

Alpacas are smart and emotionally intuitive. Alpacas can and will outsmart you if you let them. In a small herd they can open gates and steal hats, but they can also sense if you're having a bad day and stay close to comfort you. In a larger herd they can just go about their business and rarely interact with humans. They adapt to their environment and their caregivers.

There is a lot of misinformation about raising alpacas. Facebook groups are notorious for providing incorrect information from people who mean well. On the flip side, Facebook groups include people who will jump on a call or rush to your farm to help in an emergency. Although alpaca owners are spread across the world, we have a close and caring community. The key to success is figuring out who is your virtual mentor and who will quickly lead you astray.

Alpacas are a commitment, so you can't randomly leave for an extended vacation. A cat can fend for itself for a few days and a dog can go to a family or friends for a sleepover. Alpacas will require someone to come on-site to care for them. The good news is if your herd is small, chores are quick and alpacas are adorable, so finding help is easy.

Alpacas are livestock animals, and just like any other livestock, you'll experience loss. This is the hardest part for me. On our farm, alpacas are part of our family, and we treat them like our family pet. When an alpaca is terminally ill, I sit with them, cry for them, and mourn them just like I would if I lost my dog. When Reba lost her baby, I cried through the delivery, and I mourned that baby with her. Because I am so emotionally bonded with my herd, loss is difficult for me, and I need to remind myself of all the good to help offset the bad.

The Ideal Alpaca Owner

So, with those pros and cons, who is the ideal alpaca owner? My short answer would be someone who will appreciate alpacas for all their goodness. And there is a lot of goodness.

On my worst day I can walk into my paddock and have all the worries of the world melt away. This is what my life needs, and this is why we raise alpacas.

More specifically, you are the ideal alpaca owner if:

- You love and connect with animals.

- You appreciate the curiosity of cats and the loyalty of dogs.

- You appreciate smart animals with unique personalities.

- You can and will take the time to learn alpaca behavior and use this knowledge to improve your interactions with your herd.

- You can and will take the time to learn about alpaca husbandry and care.

- You have the land and shelter available to create a healthy environment for them.

- If you're breeding, you have a vet, university, or mentor available to help with medical issues.

- You can physically clean up poop, deliver hay, and trim toes—or you have hired help to do so.

- You have the base investment available to purchase a healthy starter herd.

I love everything about alpacas, and I appreciate every bit of their unique, quirky, and loving personalities. They are my everything. But that said, I've invested in learning about them, caring for them, and building a business around them.

As alpaca owners, we receive what we contribute to the relationship. The beautiful thing is there are many different types of relationships, and you can decide which style works best for you and your farm.

CHAPTER 2

TOP CONSIDERATIONS AND DECISION POINTS

Specific Factors to Consider

As you begin to contemplate what you want your future alpaca farm to look like, let me provide some key elements for you to consider. These should help you think through your options and evaluate what might work best for the lifestyle you envision.

While you won't immediately have answers for all of these items, they are data points you need to think about as you embark on your journey into alpaca farming.

Huacaya vs. Suri

There are two types of alpacas: Huacaya (pronounced wah-KI-ah) and Suri (SOO-ree).

Huacaya alpacas account for over 80 percent of the United States' alpaca population and over 90 percent of the Peruvian alpaca population. They are easily located and readily available for purchase. Huacaya alpacas look like giant teddy bears, and their fiber is fluffy and buttery soft.

The less common type is the Suri alpaca, which has long, silky fleece. Due to the low availability of Suri alpacas, they are more difficult to locate, and while their fiber is absolutely beautiful, it is harder to find a mill to convert it to yarn.

Most alpaca farms will have either Huacaya *or* Suri alpacas; few farms will offer both.

Huacaya alpaca

Suri alpaca
PHOTO PROVIDED BY AMBER STOKOSA
AT GREEN ROOM ONE, LLC

Rescue, Pet, Fiber, or Show

When starting your alpaca farm, you'll need to locate a starter herd. You have the choice of rescues, pet quality, fiber (hobby) farm quality, or high-end show quality.

While rescue alpacas are generally free, they do have their limitations and risks. Rescue alpacas have an unknown pedigree, and a lot of times their medical care was limited. This can lead to inbreeding, which in turn can bring a lot of unexpected medical issues.

Pet alpacas will be inexpensive, although they will not offer quality fiber for producing yarn or birthing crias (baby alpacas). My husband refers to pet alpacas as "lawn ornaments" because their only value is love and entertainment. For some, this might be exactly what you're looking for.

A fiber-quality alpaca (aka hobby farm alpaca) will be higher quality than a pet, but lower quality than a show alpaca. They will have strong fiber characteristics that will lead to the production of high-quality products and seed stock.

Show alpacas are the most expensive, but these will come with great fiber, solid conformation, and excellent genetics. If you are purchasing a show stock herd, look for show wins and tangible data on the fiber itself. Expect to pay a much higher price for this herd.

Some farms will specialize in just one level of alpaca, while others (like our farm) will opt to offer a variety.

Boys vs. Girls

Many new farms opt to start their farm with boys or girls rather than a combination of both genders. I'll be honest about the fact that there is no perfect gender, and it really depends on budget and personal preference. (And remember, male and female alpacas must be kept in separate enclosures!)

Teddy and Levi were best friends and always playing.

Our boys romp around the paddock playing and roughhousing. The younger boys play, and they don't fight; however, the breeder males will have distinct arguments until their social hierarchy is well established. We spend many nights watching from the house as the boys entertain us with their juvenile antics. I equate it to watching human boys in middle school. They play hard while still being great friends. If there is a disagreement, it ends quickly and they move on. That said, boys do not have the same level of drama you'll find with females.

Our young girls are sweet, silly, and get the "zoomies" where they run around wildly or prance through the paddock in a parade-style gallop. Unlike the younger boys who play-fight, the adult girls will have full disagreements. They will argue, spit, and fight for dominance until one of them finally backs down and walks away. It doesn't last long, but it does happen.

Our girls and boys both include a mix of introverts and extraverts, as well as mild and strong personalities. Both sexes will have best friends and enemies they prefer to avoid. We love our boys and girls equally, so either option is a sound one for us.

When budgeting for your alpaca purchase, it is important to know that the industry only uses the top males for breeding; therefore, we have an excess of nonbreeding males available in the United States. These males will make excellent pets and fiber producers, while also offering a lot more value for your budget.

Young vs. Old

The average alpaca lives to be in their late teens, and some live well over twenty years. This gives you lots of options when considering alpaca age.

Young alpacas will live longer, and you'll have more of an opportunity to imprint on them, but they will also be more expensive. Older alpacas will be more set in their ways, but they are also very relaxed and make lovely companions.

The decision for age should be dictated by your budget, intended fiber usage, and whether or not you desire to birth crias.

Fleece Usage

I always ask new alpaca farmers how they plan to use their fiber. Their answer will help decide the alpaca quality that is needed.

If you'd like to produce fiber for yarn and clothing, you'll need an alpaca that has fiber under thirty microns, and you'll need fiber length of three inches or more. If you're crafty and felt is more your style, an older or lesser-quality alpaca will do just fine. Knowing how you'd like to use your fiber harvest will help you determine the quality and age of the alpaca you'd need to purchase.

Herd Size

Just like gender, there is no perfect herd size. We've already covered the industry benchmark for owning a minimum of three or more, but how far you go up in numbers from there is up to you.

We started our herd with five alpacas so we could begin our farm with a number large enough to produce a healthy environment but small enough to accommodate the alpacas in the paddock space we could quickly create.

The starter herd you purchase should align with your available space and budget. Keep those data points in mind and let your farm visits provide additional direction.

Environment and Weather

The area you are in can have an impact on whether you should own alpacas and what type of alpacas you should own. Alpacas do very well in a variety of elevations and temperatures; however, they do not do well in extreme heat. If you live in a very hot climate and want to raise alpacas, you'll need to be prepared with plenty of fans and misters throughout the summer months. And while Huacaya alpacas can do well in cold weather, Suri alpacas won't do well in extreme cold.

Available Acres

Alpacas will need ample space to roam, and they'll need a solid source of pasture grass or hay. If the alpacas are going to free range on pasture, plan for one acre for two to eight alpacas. If this isn't available, make sure you have a solid hay source ready.

Zoning Restrictions

Know your zoning for raising livestock, building barns, and opening farm stores or operating an agritourism business, if this is in your plans. Remember that alpacas are categorized as livestock, which means they fall under agricultural zoning.

Budget

This is the hardest consideration for me to manage. I set a budget for an alpaca purchase, then I fall in love with a specific alpaca or lineage and my budget falls apart. My favorite girl, Nibbler, is a result of me ignoring my budget because I fell in love with her silly personality.

Keep an Open Heart and Mind

If you are looking at raising alpacas for the first time, pick an alpaca breeder who will take the time to help you find your perfect match. This could be a female or male alpaca, it could be an outgoing or reserved alpaca, or it could just be a rockstar breeder.

Go into the process with an open mind and an open heart. Visit farms, talk to alpaca farmers, and find a mentor who will help you get started and be there to support you along the way.

CHAPTER 3

STEPS FOR STARTING AN ALPACA BUSINESS

For readers interested in starting an alpaca farm *and* running an alpaca-based business, this chapter is for you. If an alpaca business isn't on your agenda, you can skip this chapter entirely.

I consider our farm to be small but mighty, and I want to share our business secrets with you. I hope you can take what we've learned and use it to turn your alpaca farm into the business of your dreams!

Planning Is the Key to Business Success

I launched my first business venture back in 2009, right in the middle of a major recession. I will never forget sitting inside my CPA's office and feeling slightly overwhelmed with everything that was in front of me. I asked her a question about future activities, and she told me not to think that far in advance because most small businesses never make it past the first year. My inner voice immediately told me it was time to find a new CPA.

My CPA underestimated me. It's well over a decade later and my first step into entrepreneurship is still going strong. I knew I could do it and I never wavered. I stumbled and made mistakes a lot along the way, but with every stumble and mistake I learned something important and grew as a business owner. With every success, I celebrated and rejoiced. And with each year, I've been incredibly thankful.

Why was I successful? It wasn't because I had superior intelligence, because I don't and my grades in school will prove it. I was successful because I took the time to do research, create a plan, and methodically execute that plan.

Success typically doesn't just happen. It's the person who makes it happen. And if you ask me, success is a direct result of research and planning. That's why having a business plan is a critical step if you want to make a profit as an alpaca owner.

When I fell in love with alpacas, my husband was super skeptical at the viability of it as a business. It took time for me to get him on board with the idea. But I did and I'm very grateful.

Before Jason would buy into owning alpacas for profit, I needed to show him why I thought we could be successful. How did I know? I did research, and I performed a Strength, Weakness, Opportunity, Threat (SWOT) analysis. A SWOT analysis is basically a document that details the strengths, weaknesses, opportunities, and threats of an endeavor. Once he saw this information, thought about it, and augmented it, he knew alpacas could be profitable if managed correctly.

Jason loves data as much as I do, and it has been our superpower. We've turned our love for data, strategy, and planning into the driving force of our alpaca business. We review it often and adapt our business based on the data we see and how the results align with our personal and business goals.

Let me take you through the steps needed to research and plan for a successful alpaca business.

13 Elements of an Alpaca Farm Business Plan

A good business plan will guide you through starting and managing your business. You'll use your business plan as a roadmap for how to structure, operate, and grow your new alpaca business. There are many different types of business plans, but I tend to lean toward the KISS (keep it simple, stupid) principle. I make sure I include everything to get started and stay on track. I also make sure I remove anything that doesn't really fit with alpaca farming.

As you read through these business components, I'd like you to keep one important point in mind: The alpaca farms that are making substantial profits are doing so by design. They have solid business plans that support a variety of revenue streams. They don't just sell their fiber. Instead, they offer animal sales, have farm stores, and market their goods through local farm markets or craft shows.

Following are the most important elements of my business plan and what my husband and I are using to guide us through our adventure in alpaca farming.

1. **Description:** This simply describes your business. Listing out who you are, who you will serve, and what you plan to offer in the form of goods and services helps you generate a nice overview of your farm and future activities.

2. **Mission statement:** A mission statement is a short paragraph about why a farm exists and provides an overview of the farm's overall goals. It's usually a few

sentences that provide the "why" behind the who. Our mission statement involves wanting to give back to the community around us. It's important to us, and it will be an important part of our future farming activities.

3. **Target market:** Your target market is the groups of people you'd like to serve. Or, in other words, who will you sell goods and services to? Defining your target market, understanding their needs, and knowing their wants will help you craft an offering that will resonate with them. When you connect with and serve your target market, you set yourself up for success.

4. **Competitors:** Your competitors might include local farms, national farms, big-box retailers, and virtually any entity that exists online. Once you define your mission statement and target market, you'll be able to list your competitors. Researching and knowing your competitors is an important part of creating a strategy for your business.

5. **Market analysis and opportunities:** You'll need a good understanding of the alpaca industry and your preferred target market. Competitive research will show you what other alpaca farms are doing right, and it will help you see what you can offer that exceeds the current state of your competition.

6. **Market threats:** Market threats could include the economy, competitors, technology, resources, environment, and really anything that could hamper your ability to execute your business plan. Knowing what these threats are will help you navigate around them.

7. **Differentiators:** Explain the competitive advantages that will make your alpaca farm and business a success. What sets you apart from other farms? What will you offer or do differently than these existing farms? How can you improve the alpaca industry or your local area?

8. **Types of revenue:** This pertains to the high-level ways your farm will make money. These can include:

    - *Transaction revenue:* Sales of tangible items (like products) that are usually one-time customer payments.

- *Service revenue:* Sales generated by providing services (like breeding, renting alpacas out for appearances, or performing tours) to customers that are calculated based on a one-time usage.

- *Recurring revenue:* Earnings from ongoing payments for continuing services or after-sale services to customers. This is predictable and ensures a source of future revenue.

9. **Revenue streams and income generation:** Explain how your farm will make money and generate a profit at a more detailed level. List this out and document any ideas that you've thought of during your brainstorming efforts. Alpaca farm income can be generated via a variety of methods that include:

   - Animal sales

   - Alpaca herdsire breeding

   - Alpaca boarding

   - Raw fiber sales

   - Manure sales

   - On-site farm store sales

   - Online store sales

   - Off-site farm markets and craft shows

   - Consignment sales at third-party businesses

   - Agritourism (like alpaca tours and trekking)

   - Photo shoots

   - Alpaca rentals (weddings)

   - Therapy sessions

10. **Revenue goals:** These don't have to be exact, but you should list out how much revenue and net profit you'd like to obtain from your specific revenue streams. I set

an annual amount and then break it down into monthly amounts. By knowing my revenue goals, I can better establish my priorities for purchases and operational activity.

11. **Marketing activities:** My background is in marketing, and it has become a large part of who I am. I would never have considered starting an alpaca farm if I didn't have an idea of marketing options for attracting sales and revenue. Most farms don't have this luxury, so let me give you a few options to consider for marketing your alpaca farm. These strategies can include:

    - Word-of-mouth referrals
    - 4-H clubs or FFA events
    - Local events like craft shows and farmers markets
    - Co-branded activities with local businesses
    - Listings in local directories like Google Business Profile, Apply Maps, Bing Local, Yelp, or Tripadvisor
    - Social media communities like Facebook business pages and groups, Instagram, or TikTok

12. **Major expenses:** With alpaca farming, the bulk of your expenses will include the alpaca herd, shelter, fencing, food, and ongoing care for shearing, medications, and an occasional vet visit. If you plan on converting your fiber into yarn, you'll need to add in the cost of fiber processing or plan on cleaning and spinning the fiber yourself. I'm not a crafty person, so we go the mill route.

13. **Milestones:** I don't think any business plan can be complete without setting some milestones. These are major events that must take place to fully execute your plan. Milestones could include financing, barn building and fencing, alpaca herd acquisition, building a website, creating a Facebook page, making your first sale, or farm expansion. Or if you're anything like us, it equates to the first barn and the second barn and so on.

Beyond the Business Plan

Once you've formulated your initial business plan, it's time to move into execution. Starting an alpaca farm can feel exciting or it can be utterly overwhelming. I've felt both emotions, and I would absolutely ebb and flow between them.

My advice is to tackle business setup in chunks, so you can focus on individual tasks and not the entire process, which can quickly become too much to manage. Just take one step at a time.

1. **Document your goals.** When my husband retired early and we moved north, we did so with the goal of slowing life down. As we started our farm and the business grew quickly, we were busier than ever before. We had to take a step back and evaluate what we were doing, because we had lost sight of our personal goals. Having firm goals to reference was very important to keeping our business profitable and our marriage healthy.

2. **Visit local alpaca farms.** Before you get too far ahead of yourself, take the time to visit other established alpaca farms. Don't limit yourself to local farms. If you don't have many alpaca farms around you, make the time to drive out of state to make sure you get well acquainted with the animals and have an opportunity to see the operations of a working farm. We visited three farms before making our starter herd purchase and I'm so thankful we did. It was immediately clear that one farm was worthy of our purchase and two were not even in the ballpark.

3. **Visit alpaca farm stores and craft shows.** While alpacas are adorable, their fiber is their true purpose, and it should not be overlooked. I firmly believe that for the alpaca industry to be healthy and sustainable, we have to make fiber usage a focus.

There is a lot to learn, so I highly recommend you visit alpaca stores and craft shows so you can see *and* touch products made from alpaca fiber. This will help you envision what you'd like to use your fiber for and how this usage can produce revenue.

4. **Attend a few alpaca shows.** In our first year of alpaca ownership, my husband and I drove all the way from Michigan to Pennsylvania to attend an alpaca show. We wanted to better understand what made one alpaca higher quality than another. I wanted to see the judges in action and hear their comments, I wanted to meet and touch those award-winning alpacas, and I wanted to see what products were being sold at the show. It was worth the drive. In one weekend, we learned more than I thought possible, and we made some industry connections that would help us along our journey.

5. **Research your local zoning laws and regulations.** Alpacas are livestock animals and keeping them requires agricultural zoning. This means you cannot have alpacas living in your house or a backyard within a subdivision. We live in northern Michigan and are within AG-2 zoning, which is used for residential development integrated with agricultural pursuits. This zoning allows us to build barns, create elaborate fencing, and raise livestock. If you do not know what zone you are in or what restrictions you have, contact your local zoning board.

6. **Verify farm tax deductions and benefits with your accountant or CPA.** Starting an alpaca farm will require a substantial investment. Before you start spending money, it is wise to speak with your accountant or CPA. Farm accounting is

THE WRONG ZONING CLASSIFICATION ISN'T THE END OF YOUR DREAM

If your existing home isn't currently zoned for alpacas, consider requesting a special-use permit. This permit allows land to be used in a manner that deviates from the normally accepted use in the area. To obtain a special-use permit, you'll need to approach your local zoning board to present what you plan to do and reasons why they should approve it. Do your homework and bring your business plan to the meeting so you are well prepared and ready to sell your dream!

different from other small business accounting, so you'll want to talk through your options and verify your CPA is familiar with farm accounting, livestock depreciation, applicable deductions, and depreciation schedules for things like barns and equipment.

7. **Select a farm name and verify it is available.** If you want to deduct your farm-related expenses on your income taxes, you'll need to demonstrate you are a true farm that was created to generate revenue. This will require you to create a formal business with your state government. The first step is to create a business name. You'll want to verify this business name isn't already taken in your state or protected with a national trademark, both of which you can check with an online search.

8. **Create a farm LLC and designate a registered agent.** Once you have your business name picked out, you need to register that with the state by forming a Limited Liability Company (LLC). While you can opt for a Sole Proprietorship, I would not recommend it. A Sole Proprietorship does not offer the same liability and asset protection as an LLC. To register for an LLC, you'll need to submit articles of organization with your Secretary of State (or similar office). You can do this yourself or you can hire a third party to form the LLC for you. In most of the United States, you will also need to establish a registered agent to manage the paperwork for you.

9. **Obtain an EIN.** An EIN is short for an employer identification number. You need this to obtain a business bank account, business permits and licenses, and to secure wholesale accounts. You can obtain an EIN for free from the Internal Revenue Service.

10. **Obtain a sales tax license.** Depending on what you plan on selling, you may need to collect sales tax. Sales tax can be calculated at the state, county, and city levels, and it will have varying requirements for products, services, and ancillary items like shipping charges.

11. **Create a business bank account.** A business bank account will help keep a formal separation between your business and personal financials. It will also help demonstrate this is a valid business and support you in a potential audit with the IRS.

12. **Review insurance options for liability and livestock.** High-end alpacas can be very expensive, and some farms choose to insure these animals. While this insurance is optional, liability insurance should not be. If you plan on having any visitors on your farm for agritourism activities, you need to obtain business liability insurance or have an insurance rider with an umbrella insurance policy. Insurance companies like Farm Bureau specialize in these types of situations.

13. **Prepare your barn or shelter.** Alpacas need shelter from extreme cold and heat, so you'll need to have a barn or similar shelter ready for their arrival. When we bought our first five alpacas, the farm that sold them to us was good about coaching my husband through barn setup. Thankfully, we had just built a brand-new gambrel roof barn, so all we needed to do was configure the interior for adult females and a cria pen.

14. **Install appropriate fencing to protect the alpacas from local predators.** Alpacas need fencing, but it isn't to keep them contained—it's to protect them from local predators. In our area, this means bears and coyotes. We have a no-climb fence that prevents coyotes from entering and serves as a scratching post for our lady alpacas.

15. **Locate available alpacas for sale and qualified farm owners.** Like any industry, the alpaca industry has a wide variety of farms. Some are experienced and take great care of their animals and others, well, not so much. You want to locate a farm that has healthy alpacas that are well cared for and on a regular schedule for feeding, shots, and shearing. Be prepared for the farm to qualify you as much as you are qualifying them. If they don't ask you about your own plans for shelter, protection, and herd size, it would in your best interest to find a new farm and a new set of alpacas for sale.

16. **Obtain medications.** Alpacas need to be protected from parasites. and they will require medication for sickness. Parasite control varies by area, but it is a necessity. Meningeal worm (or m-worm) can quickly kill an alpaca, so you must protect your herd if you are in an area that has white-tailed deer. Your breeder can tell you what protocol is being used and what they recommend based on your geographical area.

17. **Purchase hay, feed pellets, and minerals.** Alpacas have special dietary needs and you'll need to be prepared with specific hay, feed pellets, and free choice minerals. (We'll talk more about these options in Chapter 5.) The pellets are optional if

minerals are available; however, I recommend them because they make great treats to use for bonding with and training your alpacas. Ask your breeder what pellets your new alpacas are accustomed to and if they can recommend a local hay source. We started with a store-brand pellet, and then once my husband was educated on the topic, he created his own feed formula that we now have manufactured for us.

18. **Purchase supplies.** New farms will need to invest in supplies such as water buckets, feeding bins, and halters. These last-minute items are the basic requirements for feeding, watering, and walking your alpacas. Tractor Supply Company or Light Livestock Equipment and Supply will have everything you need.

19. **Locate a vet familiar with alpacas.** Alpaca vets are few and far between. Some large animal vets know very little about alpacas and are hesitant to care for them. We are lucky to have an experienced alpaca vet close by, and we also have access to professional vet schools at Michigan State University and Ohio State University. Plan ahead on this step, because you may have a harder time locating assistance than you expect.

20. **Locate a shearing team.** Alpaca shearing is not for the faint of heart. It's part art and part dexterity. Most alpaca owners opt for professional shearing teams because they have the knowledge and equipment needed to provide a fast job that will keep your alpacas safe and maintain the fiber quality. Plan early and locate a qualified shearing team as soon as you obtain your first alpaca herd. They'll have set schedules for moving through your area, and they will book up months in advance. Your attention to detail and proactive planning will be well worth it when springtime comes and the alpacas need their winter coats removed. Alpacas must be sheared annually and before warm weather arrives, so you cannot skip this step.

21. **Sign up with the Alpaca Owners Association.** The Alpaca Owners Association (AOA) is the world's largest alpaca association, with thousands of members and over 200,000 alpacas in its registry database. The AOA oversees an internationally recognized pedigree registry, the Expected Progeny Difference (EPD) program, and the alpaca show system within the United States. An AOA membership will allow you to access educational materials, register your alpacas, and participate in alpaca shows. Other countries will have their own version of the AOA, so check with your mentor or breeder to see what organization is equivalent for your area.

22. **Initiate the ownership transfer of your alpacas.** While pet alpacas will most likely not be registered with the AOA (or equivalent association), higher-quality livestock will be registered in the national database. You'll want to transfer these alpacas' registry to your farm once you've paid for your alpacas in full.

23. **Review business licenses and permits.** For the most part in the United States, you do not need to have a business license for an alpaca farm. If you decide to sell the manure as fertilizer, this changes. As soon as the word "fertilizer" is mentioned, the need for a business license pops up because this item falls under consumer protection laws and has enhanced scrutiny associated with it. Research what you need at your local and state government websites.

24. **Get to know your new animals.** You've done a lot to get to this point, and after you bring your alpacas home you should take the time to enjoy the moment by getting to know your herd. Making friends with your alpacas will be so important for handling them and loving them. I like to do this by blocking dedicated time to being present. Just go out and sit with them. Talk to them, let them know you are a safe person, and give them an opportunity to settle into their new home and learn about you in the process. Keep your hands and arms to yourself, so they can start to view you as a friend and not as a threat.

MAXIMIZE REVENUE

To make the most money from your business, you'll need to use as much of an alpaca's fiber as possible. You'll use everything from the prime blanket (the best part of the fiber that comes from the alpaca's back and sides) to the fur around the feet. It all has value, and it can all be made into goods for sale.

Don't Listen to People Who Have Failed

People often say the alpaca industry isn't profitable and I will refute that statement. Alpaca farms tend not to be profitable because owners fail to *run the farm like a business*. They don't create a business plan, and they don't have a strategy for reaching a specific target market and serving the needs of their buyers.

Don't ever let someone tell you there isn't money to be made in alpacas. They are wrong. Just because they failed doesn't mean you will. My husband and I are a testament to the fact that you can make money raising alpacas. We were profitable quickly, and we've continued to make money each year. My tax bill to the IRS is proof.

The takeaway is we've created our farm as a business, and we run it like a business. We have ongoing discussions about what is working, what isn't working, and how we need to pivot. This has helped us become profitable, stay profitable, and be happy while we do it.

CHAPTER 4

VISITING FARMS AND FINDING A MENTOR

One of the most important things you can do in your alpaca journey is to visit multiple farms before you purchase your starter herd. These visits will help you see how different farms operate and how different alpaca herds act—and you might even discover the perfect mentors along the way.

When potential alpaca owners visit our farm, we are usually their first farm visit. During their visit, I always recommend they visit more farms. While sending them to other farms might mean I lose a sale, I still encourage this action because I know it is best for the new alpaca owners and their future herd. Make no mistake, the more farms you visit, the more you learn, and each trip will be well worth the time spent.

In the world of alpacas, you'll discover there are many ways to do things. Some are right, some are wrong, and some are just different. Visiting multiple farms helps you see the various options available, and you'll be able to better determine which methodology aligns best for you and your future alpaca herd.

Our Initial Farm Visits

Our Random Farm Visit

Our first farm visit was done randomly, and while the owner was very gracious, I do not recommend you follow our lead. You'll want to plan out your farm visits and make sure you create the right expectations before arriving.

This farm had both males and females, although the owner would only show us the females. She told us she refused to interact with the males for fear of creating a situation called "Berserk Male Syndrome." (I'll talk more about this in Chapter 7.) After some research we realized this was a little unfounded and it wasn't an approach we would take with our farm. It worked well for her, though, which means it wasn't wrong, it was just different.

The farm had also decided not to register their alpacas with the national Alpaca Owners Association (AOA), which is something we also found strange. Registering your offspring helps you and your future buyers track lineages and genetics, which is important for breeding and showing alpacas. We knew this and because this farm didn't register their offspring, we were not inclined to purchase there. Again, the farm wasn't wrong in their actions, but they were different from what we planned for our farm.

Even though we decided not to purchase from this farm, the trip was valuable. I fell hopelessly in love with alpacas at this farm, and this kind woman was a catalyst for us building our new life as alpaca owners.

Our Scheduled Farm Visits

While I was traveling for work, my husband and his aunt visited two alpaca farms. The first one was so dirty and unkempt that my husband swore he would never take me there—and didn't.

The next farm on his list was Crystal Lake Alpaca Farm, which is located about twenty minutes from us. This farm was well established and well respected in the

industry. My husband had an opportunity to see how to correctly set up a farm, and he was able to ask questions about the business opportunities available to alpaca breeders. This visit took him from absolutely not wanting alpacas to being much more open to the idea.

Once I returned home, he scheduled a formal second visit so I could see the farm operations and look at some potential alpacas to purchase. I was smitten in every way. We ended up purchasing our first five alpacas from Stephan and his wife, Kristin. They were very helpful in answering questions as we initially got started, steering us toward the right fencing and barn setup, and giving us a list of initial items to purchase. When our first baby came, even though the dam (alpaca mother) was experienced, she was unable to deliver this cria on her own. We didn't know what to do, so we called Stephan for help. He came right over to help us deliver a healthy baby boy named Rasmus, and we were able to save his mother Ariana in the process. Today we consider Stephan and Kristin our friends, and I wouldn't hesitate to recommend a visit to their farm.

In retrospect, we should have visited a few more farms before jumping into alpaca ownership. That said, we could easily see the difference between the three farms, and we knew which one we wanted to model our future after. Stephan was professional and knowledgeable, and it was clear he cared for his alpacas. He was also running his farm as a business, which was our personal goal, and that mattered to us as well.

Make the Most of Your Visit

Before you jump into farm visits, I recommend you spend some time thinking about what type of farm you'd like to create and what kind of people you'd like to meet. This will help you focus on which farms to visit and help you ask the right questions.

When visiting farms, pay close attention to the following items:

- Is the farm professional in your communication and attempt to schedule a visit?

- Is the farm clean and well taken care of?

- Are the animals in good condition and do they seem happy?

- What is the vibe like? Do the animals want to interact with humans, or do they scatter when humans are near?

- Did the farm show you around so you could see their pastures, fencing, barns, and store (if applicable)?

- What did you see that you would like to incorporate into your future farm?

Topics to potentially discuss with the owners while visiting include:

- Farm setup and design, which includes land usage, pasture management, shelter options, and fencing

- Alpaca selection based on your needs, objectives, and goals

- Health, nutrition, and general care

- Breeding and birthing (if applicable)

- Local vets and alpaca specialists

- Revenue generation options (if applicable)

- Business plan creation and execution (if applicable)

During your visit you'll want to take plenty of notes. Bring a notebook with you to record answers and document any references provided. Take photos of things that stand out to you. And as you prepare for your visit, skip the perfume and smelly lotion. The alpacas will be much more inclined to interact if you don't smell like the perfume counter at the local department store.

The Importance of Mentors

I cannot stress the importance of finding a good mentor. If you only have a few alpacas and you are not breeding, a mentor is a luxury. If you're planning on breeding, birthing, and raising babies, your mentor will be invaluable.

A good mentor will provide emotional support, answer questions, share techniques and experiences, and offer resources and industry connections. All of these items are super important when breeding, because birthing comes with a lot of nuances that just cannot be covered in books.

As you consider possible mentors, know that personality fit is as important as knowledge. There have been plenty of people we've met who have had decades of experience raising alpacas but were not well matched for our personalities, ethics, and business goals. And on the flip side, we've met other alpaca breeders with whom we've instantly connected.

When looking for a mentor, consider the following sources:

- Local farms where you'd potentially buy your starter herd
- National and regional industry associations
- Alpaca shows
- 4-H and Future Farmers of America (FFA) events

Our Mentors Kim and Nancy

Early in our alpaca journey, we happened to run into two lovely ladies named Kim and Nancy from Fun in the Country Alpacas in southeast Michigan. They had been breeding alpacas for decades, and there was just something about them that drew us in. Kim was a physician who oozed compassion and could provide tons of information in bite-size, digestible pieces. Nancy was very experienced and knowledgeable like Kim, but she presented information in a matter-of-fact way that spoke to my soul. I could also tell the alpacas truly loved Nancy, and this meant the world to me. I was immediately drawn to these two ladies, and I knew they were "my people" in every way.

Over the years we have bought over twenty alpacas from Kim and Nancy, and they have become our trusted and very valued mentors. When things go wrong, Nancy is our first call, and what truly matters is she always answers. She'll not only have the answer we need, but she'll also usually deliver it with humor that makes us laugh in even the most stressful situations. And the best part is Nancy doesn't try to be funny; it is just the way she is, and we love her for it.

One important thing to note is our relationship is not one-sided. We have purchased animals from them and sold animals for them, we've delivered animals for them, and we've sold their products in our store. In return Nancy and Kim have kept us sane, Nancy has driven hours to come shear our babies, they've taught us handling techniques and provided invaluable expertise on cria health, and helped save failing crias (not once but twice), and they've hosted us at their farm for meals and fellowship.

Our relationship with Kim and Nancy is built on respect, admiration, and friendship. And while there is monetary value that flows each way, the relationship is so much more.

Our Mentor Lynn

Lynn Edens from Little Creek Farm Alpacas in New York (aka Snowmass Alpacas) is another alpaca breeder we have come to rely on. But unlike Kim and Nancy, she resides in another state, and she has a very large farm that focuses on producing high-end seed stock.

Lynn's business model isn't anything like ours, so one would think she wouldn't make a great mentor for us. But that would be an incorrect assumption.

Lynn loves data and she uses data to guide her decisions. Jason and I do the same, and have throughout our professional careers. The moment we interacted with Lynn and could see how she used data in her business, she captured our attention, and we knew she was someone we wanted to interact with and learn from. She is highly intelligent and a smart businesswoman, she has a wicked good memory, and she wants to do good for her alpacas, her customers, and the industry. All those things resonated with me and my husband.

Over the years we've purchased many alpacas from Lynn, but I would certainly say our relationship isn't financially based. She has provided breeding guidance and business advice, and she has offered history and the backstory for the alpaca industry that we wouldn't and couldn't find elsewhere. Lynn has an amazing ability to provide clarity in all things alpaca.

Relationships Are Invaluable

People are often surprised at how much we know about alpacas, how established our business is, and that we were profitable soon after starting our business. My response to such comments is that we've had wonderful people helping us along our journey.

Our success began with our first interactions with Stephan, the moment we met Kim and Nancy, and the first time Lynn wowed us with her interpretation and use of data. With each moment we became sponges that eagerly soaked up as much information as fast as we possibly could, while at the same time trying to be conscious of not creating a one-sided relationship.

It is the purposeful connections and the willingness to listen that has helped us create a successful business and a healthy, happy home for our alpaca herd.

Make no mistake, we are where we are today because we've had amazing help to get us here. And that, my friends, is the power of visiting multiple farms and finding a mentorship relationship that works for all parties involved.

No Mentor Is Not the End of the World

While I have just emphasized the importance of mentorship, I would also like to say the absence of a mentor should not deter you.

If you can find a good mentor, great. The relationship will serve you well. If you cannot, the world won't end, and you can still have a healthy and happy alpaca herd by using other industry resources.

CHAPTER 5

Creating the Ideal Farm Setup and Infrastructure

Minimum Standards of Care for Alpacas

Before you bring any alpacas home, you need to make sure you can cover the minimum standard of care for housing your new herd. There is more involved here than people think.

Since alpacas are much different from other livestock animals, there are a few basic things all new owners should know. We typically cover these items and discuss them in depth when hosting new owners to our farm because they are important for the physical and mental well-being of your herd. A few of these I've already mentioned, but they are worth repeating.

You should have at least three alpacas of the same sex. We sometimes will recommend five as a starter number depending on the alpacas being rehomed. Alpacas are herd animals, so they need a herd to feel physically safe. If they don't have a herd, their mental health suffers, and this quickly leads to a physical decline.

If you're new to alpacas this requirement can sound a little silly, but trust me when I say it is not. Alpacas are intelligent animals, and they have a lot of awareness of their surroundings, which includes other alpacas, predators, and humans. They feel much safer in a group.

An illustration of this on our farm is when an alpaca named Skyfall was sold to a lovely home with good people. Skyfall went to his new home with three other male alpacas, which was well within the minimum standards of care. The farm was set up perfectly, he was well loved, and everything was as it should be. However, once Skyfall settled into his new home, he began to have behavioral issues. He would spit at the humans during chores or any other interactions. After medical issues were ruled out, we brought Skyfall back to our farm so we could assess his behavior. And guess what? The spitting went away entirely. Skyfall returned to the sweet boy he was and spent the entire summer entertaining guests during tours. I had suspected that Skyfall didn't feel safe in his new, smaller herd, and I wanted to see if returning to a larger herd would make him less anxious. It did—he was happy, and there were zero issues with spitting. It was clear that Skyfall's herd needed to be much larger than the normal alpaca.

Now keep in mind that Skyfall is an anomaly, and most alpacas are perfectly happy with three to five alpacas in their herd. I provided that example so you can see that herd size does matter and it can impact mental and physical outcomes.

Males and females cannot live together in the same pen or barn. Female alpacas do not have a traditional cycle, so they can mate at any time. The males know this, and their desire to breed is present 365 days a year, 7 days a week, 24 hours a day. If males and females are together, breeding will happen, and most intercourse occurs at night when the humans are fast asleep.

Should frequent breeding occur, it will compromise the female's health by tearing her apart internally. This will eventually kill her if you leave the male and female in an enclosure together.

This outcome also applies to gelded males. Like a human, gelding an alpaca takes away the ability to produce offspring, but it does not take away the ability to have intercourse.

Breeding farms will need additional shelter and paddock areas for small boys. If you plan on breeding alpacas, it is important that you reserve shelter and paddock space for young males.

At roughly six months old, alpaca males will be old enough to potentially breed a female, so crias are traditionally weaned at this time. These young males are still too young to go into the regular adult male area, so they will need to be segregated until they are about two years old. Your farm setup needs to include this additional space.

Should you ignore this advice and place young boys in with adult males, the older males will mount and attempt to breed the younger males. The young males will sustain internal damage, and in many cases also external physical damage (or death) from the weight of the older male.

Alpacas need space to move. Alpacas cannot live in a traditional backyard; they need an area big enough to roam and play. An alpaca generally doesn't run around a lot, but they are known to pronk (run and skip) occasionally, and young alpacas love to wrestle. You need to make sure they have enough room for this type of activity. If you don't have adequate space, you'll create unnecessary stress on the alpacas, and you'll have to deal with a whole lot of fighting.

Alpacas need shelter from the elements. The minimum standard of care for an alpaca in the United States includes shelter from weather. A barn or three-sided shelter that will protect them from extreme cold, heat, or rain is needed.

Alpacas will need electricity in extreme weather. Alpacas are not high-maintenance animals, but they do need water in winter, which may require a heated water bucket to prevent freezing in cold weather. In warmer months they'll need fans to keep cool. Even in northern Michigan, we have fans running in every barn throughout the summer so the herd can get out of the sun and cool off in front of a fan when desired.

Alpacas need proper fencing. Alpacas are prey animals, so they have virtually no defense capabilities. Fencing is designed more to protect alpacas from predators than

accommodate this, build interior sections with wood and screws. This will allow you to quickly reconfigure the barn to meet the needs of your current state.

Fencing and Gates

I touched on fencing above and highlighted the need for taller fencing to keep predators out. Again, a five-foot, high-tensile fence with a "no-climb" pattern works great. It keeps the coyotes and other predators out but is also safe for the alpacas.

There are a lot of fencing options available, and as you browse through the various alternatives, remember to avoid electric and barbed-wire fencing, as well as anything that has wide enough gaps to allow predators in or alpacas to get themselves stuck. There are times where alpacas are like large cats, and trouble can just find them. On our farm, Auburn and Xavier are notorious for sticking their heads through the fence, getting stuck, then being forced to wait for one of us to discover the issue and wiggle them out. The good news is alpacas tend to outgrow this behavior as they mature.

Gate selection will follow the same guidelines as the fence itself. The one addition is to carefully consider your gate closures. If you are having large agritourism activities, you may want to have latches that allow for easy locking. This is needed when you have an open farm day with thousands of visitors arriving. A lot of visitors won't understand the nuances of farm life and safety, so you'll need to protect humans and alpacas alike. Having a gate you can lock with a key or combination will save you lots of hassles as visitors come and go.

Alpacas are curious like cats, which means they can and will find trouble.

Paddock Areas

A paddock area is a smaller area where your alpacas reside when they are close to the barn. This is different from the larger pasture area where they graze.

We have paddocks outside each of our barns, and we make sure they are big enough that the alpacas can play around and get exercise. The maternity paddock has additional space so the babies can run around and burn off their energy.

The fencing for the paddock needs to be very secure since this is where the alpacas will be when not out grazing in a pasture. You'll want to make sure this area is also predator proof and that the alpacas are free to come and go from the barn as needed.

Pasture Areas

Pasture areas are designed for grazing. Many farms will have a massive pasture area and then subdivide this for girls, boys, or different age groups. The benchmark for spacing is up to ten alpacas per grazing acre (this can be less if you're supplementing with hay). Due to parasite concerns, you don't want your alpacas sharing pasture areas with other livestock.

When selecting your pasture grass, remember that not all grass is ideal for alpacas. Orchard, Timothy, brome, Bermuda, and bluegrass are all suitable options. Ryegrass, Kentucky 31 Tall Fescue, clover, and alfalfa should be avoided. Rye contains a fungus that creates staggers (muscle tremors and incoordination), fescue can contain fungus and toxins, clover can upset the stomach, and alfalfa induces weight gain. I highly suggest you check with your local agricultural professionals, since pasture needs vary based on geography and weather.

Another consideration when planning your pasture area is toxic plants and trees. While there is a larger list of items to be careful of, the most common plants of concern in the United States are azalea, beggar-ticks, burdock, foxtail, hemlock, milkweed, mushrooms, poison ivy, rhododendron, thistle, and wisteria.

Finally, you'll want to carefully walk your pasture area to look for holes, plan the layout to allow for frequent manure cleanup, and make sure the alpacas have access to shade and water while grazing.

Cameras

When we first moved into our farm and were setting up our utilities, we had the internet company out to run the cable to our new-construction house. I asked the engineer about running power to the barn, and he laughed and told me we were no longer living in the city. He reminded me that I was now living in the north and most farmers don't have internet running to their livestock barns.

We listened to him, but only short-term. As soon as we started to have babies, the cameras became a necessity for us. Why are cameras so valuable for breeding and birthing? They allow you to watch from afar and at all hours of the day and night. Our cameras feed to apps on our cell phones, which is ideal for quick check-ins. You can see natural alpaca behavior without interrupting the herd. This means you'll be able to notice an illness quicker, you'll have a much better idea of labor approaching, and you'll be able to see if that new momma is letting her cria milk.

Our Google Nest and Ring cameras have been worth their weight in gold! And to be honest, they have saved lives—and our marriage. Instead of me hyper-focusing on something and annoying my husband, I can sit and watch my phone to keep my sanity in check and our marriage peaceful.

Nightlights

Alpacas don't want to sleep in full light, and they don't want to sleep in pitch dark. They enjoy a soft nightlight to offer a little illumination throughout the night. All of our barns have these available and in use.

If you want to be really progressive, you can use automation to set up routines to have your regular lights come on and off at certain times and your nightlights at alternate times. Since I am a former city girl, I will admit I set this up on Google. It makes life easier for me and my herd.

Additional Barn Supplies

As you plan your barns and layout, don't forget about purchasing and arranging things like hay feeders, water buckets, feed buckets for grain, and a bucket for loose minerals.

Alpacas are cousins to the camel, so they don't drink an overabundance of water. The average alpaca will drink about 1–1.5 gallons (or 5–7 quarts) a day, and this needs to be clean water. Our girl barn has automatic waterers that are built into the cement. These work amazingly well for all types of weather. Our boy barns have small automatic waterers in spring through fall, then we fall back on heated water buckets in winter. There is a significant price difference between all-weather waterers and seasonal waterers. The fancy, built-in waterers (Jug) are $1,000-plus each, while the seasonal variety (Little Giant) is about $50 each.

Our feed buckets are simple $10 buckets that can be bought from any farm supply store. They have a lip so they fit over a railing and can easily be moved around. These are lined up inside our barns, one for each alpaca. Each morning and night we use these to provide grain. We do separate feeding buckets to reduce arguments and to allow everyone to have sufficient time to eat. You'll want to place the buckets low enough so the alpacas can naturally bend down to eat, but not so low that poop can collect in them. We have an additional bucket in each barn to hold loose minerals, which provide vitamins and supplements. These minerals are specifically formulated for alpacas, who will eat them when they know their bodies are in need.

Our hay bins are made of metal and wood. We purchased two from a farm we helped and then my husband built the rest. We make sure we have hay feeders both inside the barn and outside in the paddock areas. Alpacas like to

Jug All-Weather Automatic Waterer

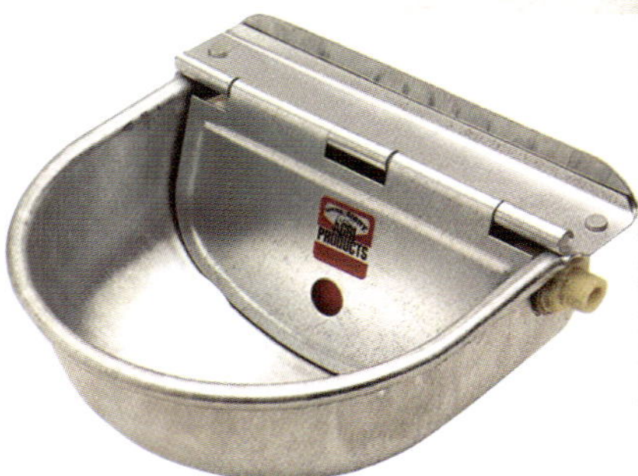

Little Giant Automatic Waterer

be outside, so if the weather is suitable, they will spend more time eating outside than they will inside. Some alpaca owners offer unlimited hay, while others provide specific amounts at regular intervals throughout the day. What, when, and how much you feed will vary based on a variety of factors such as your herd age, hay quality, and weather. Talk with your mentor or other local alpaca owners about best practices in your region.

Large metal garbage cans work great for storing grain close to where you'll need it while also keeping curious alpacas out of the food so it doesn't become a free-for-all buffet. Trust me when I say you will need these; they will make your life easier. Alpacas are smart, and they will get into things if they are not childproof.

Our Farm Layout

I believe the easiest way to learn is to have tangible examples, so let's look at our setup at Cotton Creek Farms. As I walk you through what we have and why, remember we have an active farm with a storefront and interactive tours, and we breed alpacas. This means we need a lot of space and buildings to keep up with the farm activity. Each farm is unique, and you'll want to adapt your layout to accommodate your land, any existing structures, and your activities.

We purchased raw land when we decided to move north. Our farm is twenty-seven acres that includes flat land, woods, a small creek, and a small pond. The diagram shown here is not all twenty-seven acres and only represents the part of the land that is used for the alpacas.

Before our house was even ready for occupancy, we built the large red barn that is labeled the "boy barn." We housed our first five lady alpacas in this barn, and it was immaculate. My husband configured the inside to be a palace for our ladies—we even made a bathroom area for them.

As our farm activities matured, we brought in two small run-in-barns to help house alpacas as we grew. We were still busting at the seams and we wanted to have a storefront, so we built the large gray barn next.

This new farm layout gave us enough space to house males and females in separate barns with a large area in between. They can still see each other, but that space in between makes a huge difference in limiting romantic interests and reducing male fighting.

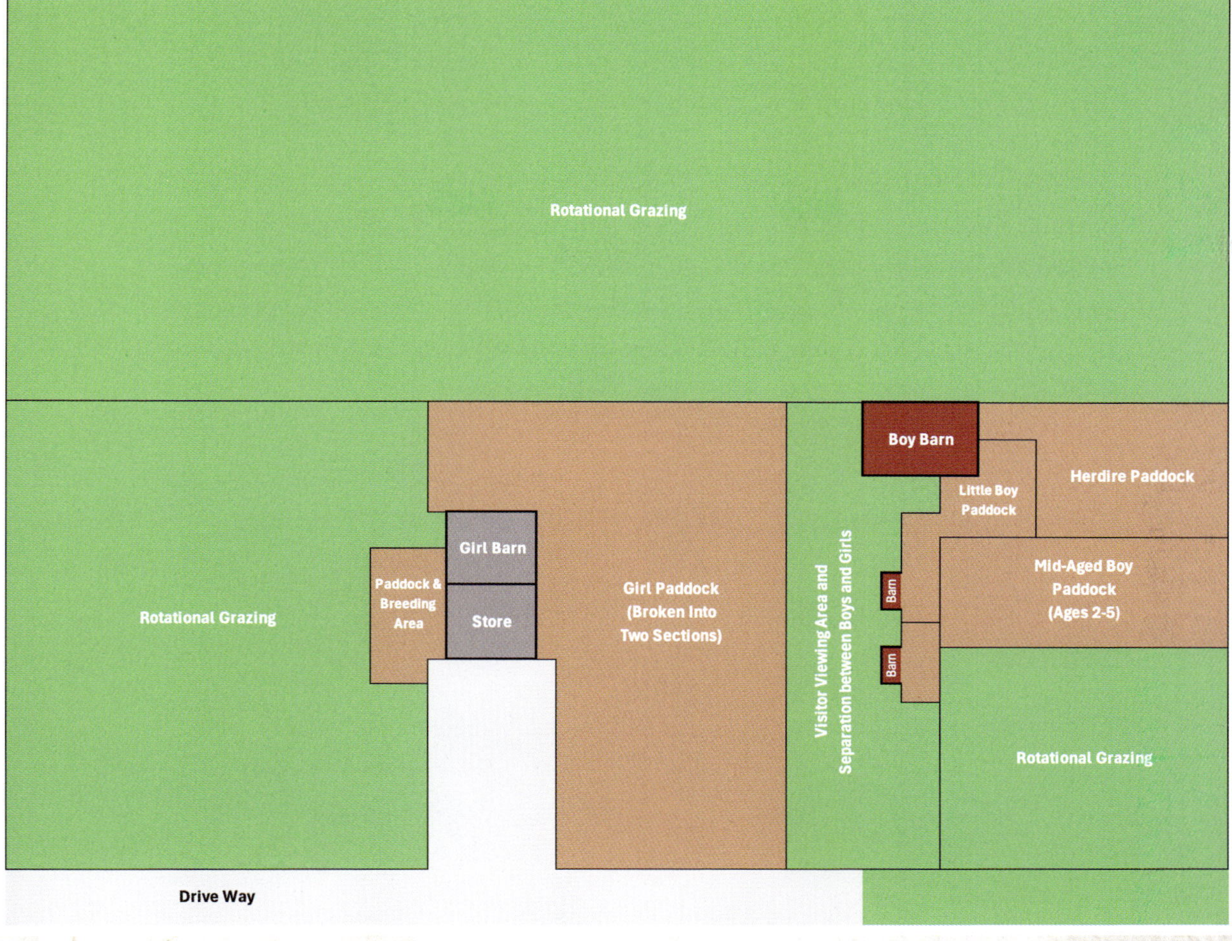

Cotton Creek Farm's layout

The herdsires (breeding males) have a section of the large red barn, while the two small barns are used for our younger males. One small barn keeps our boys right after weaning, and the other one is for our 4-H boys who are gelded and much more mellow than the herdsires.

The large gray barn is split into two main sections, with the front housing the store and the back housing the ladies. The ladies have their area split into two sections so we can separate the maternity and nonbreeding ladies. We do this so we can feed extra grain during pregnancy and post-delivery without creating weight problems with the regular girls. The back portion of the gray barn also has a medical room, a storage area, and a cria pen, which allows the babies to come in and eat grain when they wish.

This cria area is also used as a bonding area for mother and baby in the first forty-eight hours after delivery.

You'll notice there are green and brown areas. The green coloring highlights pastures that are used for grazing and the brown coloring shows areas that are dry paddocks. My husband hates the look of the dry paddocks, but they offer protection against parasites. We have from five thousand to ten thousand people come through our farm each year, and when we do tours we keep the humans within the

dry paddocks. This helps stop any parasites entering our farm via their footwear, and because we use this approach, we have had virtually no issues with parasites.

In the aerial view of our farm, you can see the diagram in application. This photo was taken on a day when the store was open and a tour was taking place.

What you are not seeing in the photo is a lot of manure. We clean up poop twice a day in spring through fall. We use a wagon and a pasture vacuum (yes, this exists)

to relocate the manure behind the herdsire area. Flies will gravitate to the manure, so you want to keep this as far away from the activity as possible.

In the photo you'll see darker areas within the paddocks. These are the communal dung piles (aka poop piles) the alpacas naturally create. The males are much better than females at keeping their piles small, and I've been told this is because pregnant females tend to spread out their urine as a way to inform males they are not interested.

Visitors often remark about the cleanliness of our farm, and while we'd like to take all the credit, the alpacas deserve the true praise. Their communal piles for urine and manure make cleanup a lot faster than with other livestock animals.

Hopefully, this walk-through of our farm setup helps you envision what you'd like your future farm to look like. Just remember, there are many ways to design your farm, and you need to configure your infrastructure to work with your land and any existing shelters, and satisfy your individual needs.

Creating Your Shopping List

For some people the planning phase of the alpaca journey is exciting and for others it can be overwhelming. I'm the type of person who loves the planning part, but that is because I love creating lists. I want to wrap up this discussion with an easy-to-use checklist that you can refer back to as you move forward. Use this as a starting point, and talk to your mentor about what else might apply.

1. **Shelter:** Three-sided at a minimum, with a minimum of eighteen square feet per alpaca

2. **Fencing:** Secure five-foot fencing for paddock areas

3. **Gates:** Easy-to-use closures with optional locks

4. **Hay bins:** One for each barn and paddock area

5. **Water buckets:** One per barn with the ability to heat in colder temperatures

6. **Fans:** Used in warmer weather to keep alpacas cool

7. **Hay:** We feed second-cutting hay (we love orchard grass) in round or square bales with enough to cover 2.5 pounds per day per alpaca for days not on pasture.

8. **Grain:** We feed pelleted alpaca feed, with enough to feed 0.5 pound per alpaca per day.

9. **Minerals:** Free choice minerals made especially for alpacas

10. **Feed buckets:** One per alpaca if feeding grain and one per barn for minerals

11. **Straw:** Great for providing bedding in extreme weather

12. **Livestock scale:** Desirable for monthly herd health evaluation

13. **Halter and lead:** Desirable for medical care, transportation, and/or country walks

14. **Rakes and shovels:** Used for manure cleanup (cheap rakes seem to work best)

15. **Wagon:** Used for transporting manure during cleanup

16. **Nail trimmers:** Used for occasional toe trimming in between shearing

17. **Hair clippers:** Used to keep fur out of eyes (not all alpacas will require this)

18. **Ivermectin or Dectomax:** Used to proactively treat for m-worm in white-tailed deer areas

19. **Syringes:** Used to administer monthly injections

20. **Fecal testing equipment:** Optional device used to perform on-site fecal tests

Remember, not every alpaca farm will require all these items. For example, we didn't start with our own fecal testing equipment, but we did purchase this later because my husband really wanted to be able to perform his own tests. Most farms provide fecal samples to their livestock vet for testing.

The important takeaway is that you want to be adequately prepared so you can welcome your alpaca herd home and enjoy the moment. In a perfect world, you'll have all your shopping done in advance so you can spend quality time with your alpacas and enjoy their rich and unique personalities.

PHOTO PROVIDED BY
LAUREN ALYSSA PHOTOGRAPHY

CHAPTER 6

THE BEGINNER'S GUIDE TO PURCHASING A STARTER ALPACA HERD

The most exciting step in setting up your alpaca farm is to purchase a starter alpaca herd. This could be three, ten, or twenty alpacas. Regardless of the number, it's an important step and one I'd like to help you with. In this chapter, I'll discuss the key things you need to know about buying your first set of alpacas.

Important Things to Do Before You Buy Your Alpaca Herd

Let's take a moment to briefly touch on some of the things you should do *before* buying. This isn't a huge list, but it is an important one.

Do Your Research

Buy books, go to alpaca shows, or attend local community events like 4-H or FFA competitions. Watch the alpacas, and learn as much as you can about them so you are in a good position to decide if a given alpaca or herd is right for your farm.

Visit Several Alpaca Farms

As I've already discussed, it's important to visit at least several breeders so you can meet the alpaca herd and get a feel for the animals. Ask the farm owners questions about their breeding program, health care, and operations. When I see someone come for a visit with a pen and paper, I'm super happy, because I know they are serious and want to take good care of their future herd. Expect to spend a few hours at each farm.

Check Online Reviews

Take the time to search Google for online reviews of any farm you might purchase alpacas from. While not all reviews can be trusted, you can spot a real one by the depth of details and the tone of the writer. If you're considering buying alpacas from a ranch with poor reviews, think twice.

If the owners of the ranch are professionals, check reviews for the career side of their life, too. Before we bought alpacas from Kim and Nancy at Fun in the Country Alpacas, I reviewed Kim's professional reviews. She is a surgeon, and I knew I'd hear the good and the bad online if I looked. I only found "the good" about Kim, and every review raved about her amazing bedside manner and how caring she was with patients. And as I suspected, this same care and love is transferred to her alpaca herd and to us as her buyers. This mattered to me, and it was a factor in moving forward with buying over twenty alpacas from these ladies.

Choose a Breeder You Trust and Connect With

When alpaca buyers visit us, my husband and I both make ourselves available to provide information and answer questions. And I'll be honest, I'm interviewing you as

much as you are interviewing me. I need to get a good vibe from you, and you need to get a good vibe from me. If we don't have that connection, this isn't a good fit and the sale should not happen. You need someone you trust and feel comfortable with, and who will be accessible to help and answer questions after the purchase.

Alpaca Purchase Criteria

As fun as alpacas are and as easy as it is to get sucked into their cuteness, I highly encourage you to purchase with caution and with purpose. Alpacas can live for up to twenty years, so you want this purchase to be the best decision for you and your future farm.

Years ago, the suggestion was to "buy the most expensive alpacas you can afford," and many breeders still use this advice as their guide for acquiring animals. In my opinion, this statement only applies if you are purchasing a starter herd with the intention of showing and winning championships.

If you are purchasing alpacas and don't plan on competing at the show level, you should "buy what you need," because that is what will serve you and your farm best.

So why do I encourage the exact opposite of long-term breeders? Alpacas can vary drastically in price, and you can waste thousands and thousands of dollars buying more than you need. That isn't good for you, your farm's longevity, or the animal.

Let's review what you should consider when you are looking at individual alpacas to purchase. Keep in mind the weight of each criterion will vary based on the purpose of your herd.

Overall Health

Review the alpaca from head to toe and look for signs of illness, weight issues, or neurological concerns. If the alpaca is smaller than it should be or has a protruding backbone, it might be hiding an illness. If the alpaca walks funny or struggles to get up, this might be a sign of neurological issues and previous illnesses.

If you are purchasing at the higher end of the price range, one option for evaluating health is to ask an experienced alpaca vet to examine the alpaca prior to purchase. This will help validate your choice and make sure you are obtaining a healthy alpaca.

Conformation

Conformation includes traits associated with body structure and soundness of bone. These characteristics include things like bite, ears, legs, body capacity, body score, proportions of body, testicles, and genitalia.

Poor conformation can be a sign of genetic defects or poor health, so look closely at this with each alpaca. It takes many breeding cycles to improve poor conformation, so good characteristics must be one of your top criteria if you plan on breeding.

Age

While age is important for all the standard reasons, it is also important when mixing in male alpacas. You cannot place young males (up to two years old) with older breeding males because it is not physically safe for the boys. If you purchase a group of males with a wide range of ages, you might need additional infrastructure to accommodate this purchase.

Another consideration with age is how soon you'd like an alpaca to breed versus how long you'd like that alpaca to breed. We do not breed female alpacas before age two, so we often have a waiting list if you want to purchase a cria (baby). But that said, some females can breed until they are fifteen or sixteen years old, while others will need to stop around age ten. Knowing if you want to breed and how long you'd like a specific alpaca to breed will help you select the right breeding females and the proper quantity for your needs.

Phenotype

Phenotype refers to traits such as color, fiber density, crimp frequency, crimp amplitude, staple length, fineness, uniformity, and so forth. The importance of these individual characteristics will vary based on the intended use of the alpaca herd.

For example, if you are purchasing an alpaca for the show ring, you will place greater emphasis on crimp and uniformity, while for a herd used for fiber production, you'll place significant value on staple length.

Color

Some alpaca buyers have strong preferences for color, while others don't care at all. Alpaca fiber comes in a wide range of colors, from white and beige to different shades of brown, rose grey, silver grey, and black. White, beige, and fawn alpaca fiber can be easily dyed, so knitters love it, but black and silver-grey alpacas are beautiful in their own right and provide rich natural colors for yarn and clothing.

You might decide to be a light breeder or a dark breeder, or have a mix of all colors. Knowing your preference before you buy is important because you'll want to match this to your future herd's visible and background colors.

Color Genotype

Color genotype refers to the color you do not see. We can know not just the visible color, but also the genetic color of an alpaca and what color fiber that alpaca can produce in its offspring. This helps us improve breeding outcomes and increase the quality of the fiber. For example, a fawn alpaca might be a black alpaca in disguise. This means that even though the alpaca is fawn colored, it can produce black alpacas.

My husband and our friend Lynn Edens have led the pursuit of this research in the United States, and our farm is very focused on knowing an alpaca's color genotype and using it in our breeding programs. We provide this information to our buyers so they can make better decisions on purchases.

Color genotyping can be a huge asset to a breeding program, so I'll cover this in more depth in Chapter 9.

COEFFICIENT OF INBREEDING

The coefficient of inbreeding (COI) is a calculation that measures how inbred an individual alpaca is. This can be calculated for a proposed offspring by combining a potential male and female.

The AOA website has a tool for figuring out this calculation, and each registered alpaca in the AOA's database will also have a calculation based on their lineage. This can be very helpful in creating a starter herd and for making sure an individual alpaca is not inbred.

Lineage

Lineage should be considered when selecting alpacas for breeding. The alpaca population in the United States has many strong genetic lines, with some lines being prepotent for certain traits like fineness, crimp (waves within the fiber), or staple length (length of the fiber).

The Alpaca Owners Association (AOA) has complete lineage information on every registered alpaca in the United States. You should review a formal AOA certificate for each alpaca before purchase. Reviewing this data will also help you make sure lines do not overlap, so you'll have lots of options for breeding your males and females.

Histogram Data

Histogram data is a fabulous way to compare one alpaca phenotype to another. The data is independently created using specific testing equipment. This allows you to objectively compare one animal to another.

Animal Name:	Snowmass Sherry Ann	Date of Birth (Age):	21-Sep-20
Registry #:	36070794	Sample Location:	GRID
Breed:	Huacaya	Sample Date:	16-May-22
Sex:	Female	Previous Shear Date:	12-May-21
Colour:	100 - White	Total Fleece Wt (lbs):	5.2

Laboratory Data

Mean Fiber Diameter:	18.7 microns	Mean Staple Length:	97 mm
Standard Deviation:	3.6 microns	Length Standard Deviation:	5.1 mm
Coefficient of Variation:	19.3 %	Length Coefficient of Variation:	5.3 %
Spin Fineness:	18.0 microns	Mean Curvature:	61.3 deg/mm
Fibers Greater Than 30 microns:	0.9 %	SD Curvature:	25.6 deg/mm
Comfort Factor:	99.1 %	Medullated Fibers:	6.6 %

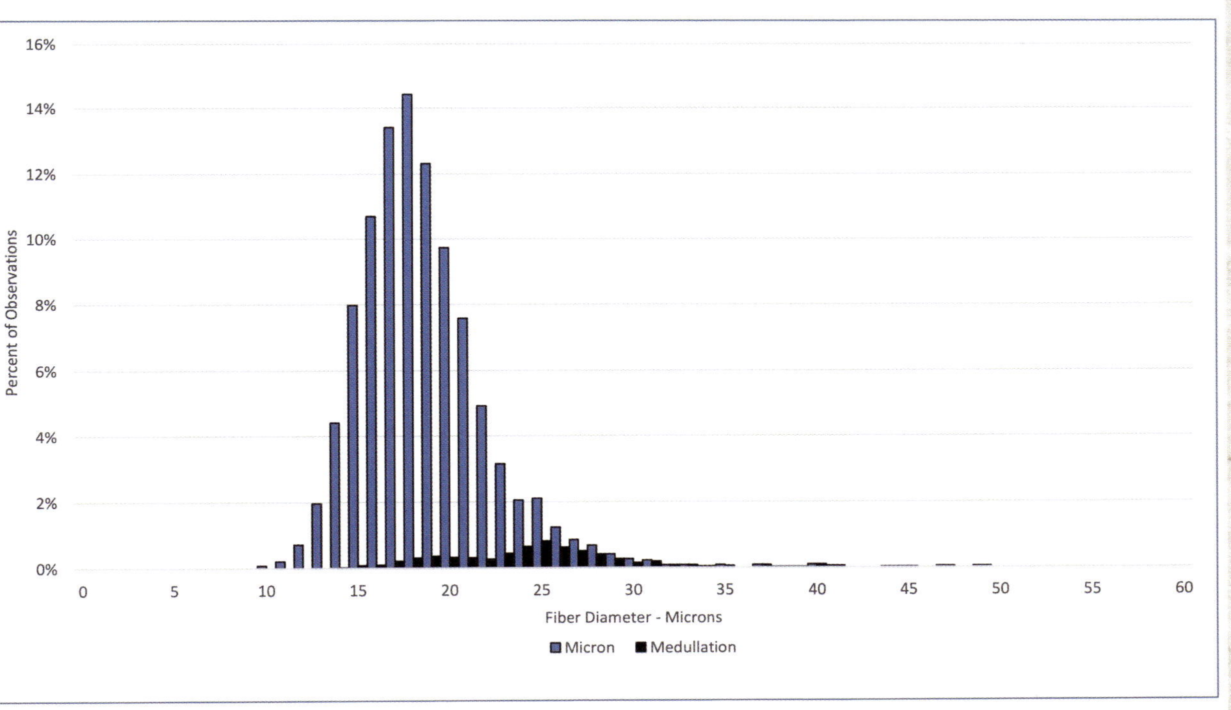

Test Performed According to Method AS/NZS 4492.5:2000

Breeding Experience

If you are looking to breed alpacas, experienced males and females can be very helpful in getting started.

An experienced female alpaca will know when and how to breed, and she will be well prepared for nursing when that baby arrives. A maiden (first-time mother) may not be as quick to drop for the male or as knowledgeable on what to do to get that cria milking.

Similarly, a proven male breeder will know exactly how to court the ladies and get them to stay still while breeding. An inexperienced male may not be able to get the ladies to cush (lay down) or might not know how to keep them still if they are a roller (yes, some ladies like to roll around while making babies).

Personality

If you care about interacting with your alpaca herd, watch the movements of the herd you visit. Are they friendly? Do the alpacas seem eager to interact with the humans? Or are they aloof or withdrawn?

While an alpaca's personality won't matter to some breeders, it will matter to someone buying an alpaca as a pet, for 4-H, or for agritourism activities.

Breeder Reputation

Choose a reputable breeder who is knowledgeable and experienced in raising alpacas. Review their website, check online reviews, and ask around to see if you can get a feel for their experience and reputation.

Select a breeder who wants to help you find the right alpaca for your needs and not just sell you any alpaca. There have been lots of situations where we don't have the right alpaca for someone, and instead of just selling them the wrong one, we use our network to locate the ideal alpaca at another farm. This shows that the breeder is focused on you and your needs, and not just on making money.

Professionalism

As you interact with alpaca farms, watch closely for signs of respect and strong ethics. Are you comfortable with this breeder? Do they offer photos? Do they offer fiber data, such as histograms (data on fiber quality) or EPD ratings (data on expected offspring fiber traits)? Will they provide a structured contract with terms and conditions of sale? If you are not receiving ample information on your purchase or answers to your questions, consider shopping elsewhere.

Mentoring

There are a lot of nuances to know about raising alpacas. While they are technically easy keepers, they are not like other animals and they do require specialized care. Make sure you buy from someone who will mentor you long-term. And by mentor, I mean someone who will pick up the phone and guide you through a delivery, illness, or business decision. These are the people you want to buy from and create a long-term relationship with.

Factors That Influence the Cost of an Alpaca

While I have already provided a lot to think about and consider, we're not done! We now need to touch on the cost of your starter herd.

The cost of a starter alpaca herd can vary widely depending on several factors, including the pedigree of the animals, their age, their breeding status, show records, histogram or EPD data, and the geographic location of the seller.

Here are some data points that will impact the cost of your alpaca herd purchase:

- **Age:** Older females will often be less expensive than younger ones; however, they will not have as many breeding years available. They are a great option if you are starting with a smaller budget or simply desire pets.

- **Show record:** Show-quality alpacas will be far more expensive than a fiber-focused herd. The more ribbons and championship banners an alpaca has, the more money the owner will want in selling it.

- **EPD and histogram data:** The better the data, the more expensive the alpaca. This is because EPDs and histograms help breeders separate the high performers from the underperformers.

- **Personality:** Friendlier alpacas cost more money. Everyone wants alpacas that are loving and want to cuddle. Since that is a developed trait and takes time, these alpacas are not easily found. That means the demand is greater than the supply, which means the price goes up.

- **Location:** There are areas in the United States where there are few alpaca farms. This means there is a limited supply, which will dictate a higher price. Don't be afraid to travel to purchase alpacas. We've purchased and sold alpacas all around the country.

- **Registration:** An unregistered alpaca will be far less valuable than one that is officially registered with the national association. This is because an unregistered alpaca has an unknown genetic history, it could be inbred, and its offspring cannot be registered.

Make a list of what matters for your purchase and stay focused on your purchase goals. My goals may be different from your goals, so follow your own list and keep a close watch on which breeders meet your requirements.

Why Do Alpacas Cost More Than Other Livestock?

Alpaca prices can seem very high to those new to the industry, but the cost structure is justified when you consider the uniqueness of birthing and raising alpacas. Supply and demand greatly influence the cost and sale price of alpacas.

Here are just a few of the nuances that make the initial cost of alpaca ownership as high as it is:

- Alpacas were first imported into the United States in 1984, so they have not been in the country all that long, which means the availability is not the same as other livestock.

- Most of the world's alpaca population lives in Peru, and the importation of alpacas from Peru is no longer allowed. Local farmers can only import alpacas from

Australia and Canada, which greatly reduces the availability of alpacas; this is especially true of high-quality alpacas.

- The alpaca reproduction cycle is not like most livestock. Their reproductive process is not conducive to artificial insemination, which makes natural alpaca pregnancies the only option.

- Alpacas are pregnant for almost an entire year, and they only have one cria per pregnancy. Twin babies occur occasionally, but it is very difficult to get both crias through a healthy delivery. Therefore, one mother can only deliver one cria per year.

- There is a natural loss of pregnancies. You can start with ten babies due and only deliver eight or nine. This is natural and needs to be taken into account when calculating cost and return on investment (ROI).

- If you do not have your own herdsire (adult male), you must pay another farm for breeding. A quality breeding will cost between $2,000 and $5,000 per cria.

- Suri alpacas have much less supply than Huacaya alpacas, so these may come at a premium price in the United States.

COST GUIDANCE NUMBERS

These prices are designed to provide cost guidance only.

Type of Alpaca	Characteristic	Male	Female
Pet	Nonbreeder Fiber not usable	$500–$1,000	$500–$1,000
Hobby/Fiber	Nonbreeder Usable fiber	$1,000–$2,000	$1,000–$2,000
Seed Stock	Breeder Usable fiber	$3,000–$10,000	$3,000–$5,000
Show	Breeder Usable fiber Show producer	$10,000–$50,000	$5,000–$20,000

Expect to Receive Contracts and Read Them Carefully

If you are purchasing a pet alpaca, it is possible you won't receive a contract. If you are purchasing an alpaca for fiber, seed stock, or show, make sure you have a solid contract and paper trail.

If you do not understand the contract, ask questions. Many contract terms can be confusing to new owners, so ask questions and keep asking them until you are comfortable.

As you review potential contracts, make sure there are enough details listed so you know exactly what you are buying and what remedies you have if you don't get what you thought you purchased.

Here is a quick list of what should be included in the contract for your alpaca purchase:

- Contract date

- Seller name and address

- Buyer name and address

- Names and AOA registration numbers for each alpaca (if US based)

- Purchase price per alpaca

- Financing options or payment terms

- Breeding options (for females)

- Breeding guarantees and/or remedies for infertility

- Delivery options or fees

- Process and timing of ownership transfer (passage of title)

- Risk of loss

- Warranties or lack thereof

- Proper care guidelines

- Jurisdiction for disputes

If you feel your contract is lacking details, pause the purchase and take the time to have the contract amended to include the missing information. Don't feel like you'll lose the opportunity to buy your dream alpaca. A good farm will hold the alpaca until you work out the details of purchase.

Beware of Alpaca Sales Scams!

While the alpaca industry is 100 percent legit, there are many online scammers mimicking our farm names, photos, and social media profiles to scam innocent and well-intentioned buyers. If you are considering purchasing an alpaca online, know it is buyer beware. You must do your due diligence or you will be scammed out of your money.

How to Spot a Scam

The typical alpaca scam will start with someone setting up a fake Facebook account or Craigslist profile. They take real photos from legit alpaca farms and use them on their Facebook pages, group posts, or in private messages.

The scammers will reach out to potential buyers via Facebook Messenger so the conversation is private. The fake seller will present an amazing offer, claiming their alpacas are super sweet, adorable babies and at a cost lower than market price. The scammer will also claim others are interested, so they'll require the buyer to transfer a deposit to hold the alpaca. Once the money is transferred, the scammer will disappear. When the buyer tries to contact the seller, the seller will block the buyer or close the Facebook account.

Sadly, Facebook has so far done absolutely nothing to help this situation or close these fake accounts, even though every time we find something hinky, we report it.

How to Avoid a Scam

To avoid falling into an alpaca sales scam, **compare the Facebook profile's URL and name.** If they don't match, it's probably a scam. Next, you'll want to carefully review the photos. Are the alpacas in different-looking barns or standing in front of different types of fencing? Since most farms will have consistent-looking styles of shelters and fencing, this is a good way of spotting a scam.

Know the market rate of alpacas and don't expect a special deal. You simply won't find a super cheap alpaca that is healthy, young, adorable, and sweet for $250 or $500. This is especially true if it is a breeding female.

Question the age of the alpaca. Quality breeders won't sell babies to brand-new farms. We tend to sell no younger than a year old, and they need to go with buddies to limit their stress.

Don't try to buy a bottle baby. Bottle feeding is the last resort to keeping an alpaca alive when the mother has either died, is unable to produce milk, or rejects the newborn. No alpaca breeder wants a bottle baby. It's a lot of work, it interrupts the cria's required nutritional intake, and it negatively impacts their socialization.

Question someone selling a "breeding pair." As we've already covered, males and females cannot live together in the same fencing and shelter. Reputable breeders know this and will never sell a breeding pair or a sister/brother combo.

Expect the seller to qualify you harder than you qualify them. A solid alpaca breeder will ask you a lot of questions because they want to make sure their alpaca is going to the right environment. I turn away about 95 percent of the inquiries we receive because the buyer hasn't done enough research or isn't prepared to raise alpacas in the environment needed. I do this because I love my alpacas and I want them to go to good homes. Remember, as the buyer, you should also ask a lot of questions about the alpaca you want to purchase.

Validate the farm with your local or national association. Is the farm a member? You also have the option of looking up the farm name in its state's business registry. A lot of farms are registered as Limited Liability Companies (LLCs), which means they will have a public record with the state. If they are a valid LLC, they'll also have an employer identification number (EIN) with the federal government.

Validate the alpaca is registered. We register all our alpacas and provide potential buyers with certificates of lineage so they can validate their potential purchase with the national association. If the alpaca is younger and we haven't registered it yet, we provide AOA registration details on the sire or dam so it can be registered in the future.

Require a contract before funds change hands. We provide a multipage contract and an invoice for all sales. This means there's a clear paper trail of ownership and payment, as well as clear expectations of the conditions of the sale.

Visit the seller's website and look for red flags. One of my favorite red flags is a scammer who publishes on a website in US dollars but lists VAT (European taxes). In the United States, we have sales tax and do not collect VAT. A true US farm would know this, but a scammer in Nigeria would not.

Finally, visit the farm prior to money changing hands. While we've sold alpacas all over the country, we love selling locally so we can meet people face to face and provide ongoing mentorship. But we've also bought alpacas from states away and have taken multiple days to drive to New York to review alpacas, spend time with the sellers, and see the farm in person prior to purchase. A good, reputable farm will welcome visits.

Some Important Takeaways

Shopping for alpacas can be an extremely fun process. You get to meet lots of amazing people and interact with lots of wonderful animals. It can be a bit of an endorphin rush!

You need to work hard at keeping those endorphins under control. Don't allow emotion to make you lose sight of your goals and the longevity of your farm. For example, if you are planning on doing agritourism, personality and phenotype are important decision criteria. Or, if you only want to use your alpacas for fiber harvesting, personality doesn't matter as much, but staple length will.

A quality breeder will ask you about your farm goals and help you select animals that meet those goals. An exceptional breeder will admit when they don't have what you need and will refer you to other reputable farms that may have exactly what you're looking for.

CHAPTER 7

UNDERSTANDING AND INFLUENCING ALPACA BEHAVIOR

I've long been a "people-watcher" who will sit back, observe body language, and pay close attention to facial expressions. This trait has helped me throughout life, but I had no idea it would also help me be a better alpaca owner and breeder.

When we brought home our first five alpacas, I would head to the barn every night and spend time with them. In freezing cold northern Michigan weather, I'd throw on snow pants and bundle up so I could go play with my new alpaca friends. In my mind I was being present to make sure the girls felt loved and to help them adjust to their new home. And while that was the case, I was also watching and learning from their behavior.

Just like people, if you stop and take a moment to observe and listen to your alpacas, you can discover a lot about them. And this knowledge will help create a happy and healthy home for them.

Does Understanding Behavior Really Matter?

Alpacas are extremely smart, and understanding their behavior is an important part of owning them. The happier your herd is, the easier time you'll have handling, breeding, and keeping them mentally healthy.

Good mental health improves their overall physical health, the interactions between alpacas and humans, and the breeding and birthing experience. It also encourages the production of high-quality fiber and helps create a better herd dynamic.

At first my husband wasn't a believer. He thought my time spent with the herd and my observations were humorous and a bit "over the top." Then he began to see how this activity impacted life on our farm. The breeding process was calmer, troubled births became nonexistent, medical issues were reduced, and our interactions with the herd were much easier and more fluid. My naysayer husband started to request my presence to check on a possible illness, help keep someone calm during a medical procedure, or move a group of alpacas from one area to another. He could see the difference in behavior when I was present, and he knew it would help create a better outcome. And honestly, all of this improvement wasn't because I have superpowers. It is simply because I've taken the time to learn, adapt, and influence our alpacas' behavior.

Alpacas Are Smarter Than Most People Realize

People are always surprised when I state alpacas are intelligent animals. And when I say intelligent, I don't just mean smart in comparison to other livestock. Alpacas are a highly intelligent animal and some also have emotional intelligence. Let's look at some examples.

Alpacas are very astute. They watch humans closely and use this information to modify their own behavior. Our alpaca herd has learned "no ma'am" and "no sir." This means they understand when they've upset me, and when they need a course correction for their behavior.

Some of our herd has emotional intelligence. They can sense my moods, and they can tell when I'm having a bad day. My favorite alpaca girls somehow know when I'm struggling, and they seem to be extra loving and attentive until my mental state bounces back to normal.

Alpacas can watch and react to the facial expressions of humans. Many of our alpacas know my "mom look" of annoyance and will quickly adjust their behavior when they see it. I don't need to make a sound; they see the change in my expression and self-correct so they can please me.

Alpacas can learn verbal commands and hand movements. There are many alpaca owners who have taught simple verbal commands to instruct an alpaca to lift a leg for toe trimming, jump in a trailer, or come into the barn. I've unintentionally taught our alpacas a hand movement to follow me, while our mentors Kim and Nancy intentionally taught their girls to run to the barn when they clap their hands.

Alpacas can recognize physical items and recall their use. We do a lot of interactive tours on our farm, which means humans want to take photos with their smartphones. These humans also bring bags of treats for the alpacas. Our herd understands the flow of a tour. If they see a human bring out a cell phone, the alpacas are very good at posing for photos, because they know the faster they pose, the quicker the treats arrive. After tours, people will show me their photos, and I cannot help but giggle because I know exactly what the alpaca pictured was doing.

Alpacas know their names when you use them frequently. This is a running joke on our farm. I've always said the alpacas know their names, and my husband has always said otherwise. The truth is the alpacas know their names, but they have selective hearing. If I call Dolly or Kalista to come out of the barn to see visitors, they'll come running. They've heard their names and know I want them to come to me, so they comply. When my husband does the same thing, nothing happens. I mean zero movement. Why? He gives monthly shots, trims toes, and performs their medical procedures. They don't like him because of it, so why would they come when he calls?

Alpacas can become jealous if they believe another alpaca is receiving more attention. This isn't true for our entire herd, but it is absolutely true for some of the smarter alpacas. Sherry Ann was born on our farm, and I've always had a strong bond with her. Nibble was not born on our farm, but I bonded with her immediately upon her arrival at six months old. Both these ladies believe I am their human; if another alpaca tries to interact with me and these ladies see it, they will subtly reposition themselves between me and the offending alpaca. They believe they should receive all of my attention, and they work to obtain it.

Alpha alpacas want to be treated with respect, and they want this respect to come from both humans and the alpaca herd. Ariana (female) and Leviticus (male) have both served as alphas on our farm. They know their social status, and they believe they deserve the most attention, the first feed at dinner time, and the most treats should apples or carrots arrive. If you fail to provide any of this, they will let you know. If you want to win over an alpha alpaca like Ariana or Leviticus, you must treat them with the respect they believe they deserve. Doing so can be as simple as acknowledging them first when you walk into the barn or giving them the first apple slice. It amounts to performing very simple actions that make them feel respected. In turn, these alpha alpacas will respect you and treat you as their alpha.

Alpacas watch and learn from others. When we've purchased or birthed alpacas, I see the new arrivals closely watching the behavior of our existing herd. They are learning what is safe, what is the norm for behavior on our farm, and what is not acceptable. Our crop of babies in 2023 were the sweetest we've ever had. I attribute part of this to my focused interaction and part to their ability to learn from the friendly adult alpacas on our farm.

Other Important Things to Know About Alpaca Behavior

Alpacas Have a Herd Mentality

Alpacas are herd animals with very limited ability to protect themselves. They feel safest in a herd, and they are at the peak of mental and physical health when they are in a group.

When our herd was at its biggest, we had just over seventy alpacas. While this was a bit large for my husband and I to manage on our own, it did give me the ability to watch and learn more about herd behavior.

During this time, we had a lot of white females and a lot of black females. Some were well-behaved and some not so much. If I were to scold a white alpaca for poor behavior, this alpaca would instantly run to wherever the other white ladies were and try to blend in with this little group. It was as if they were seeking shelter from like-colored animals so they could blend in better.

I've also noticed alpacas will nap in distinct color groups—blacks with blacks and whites with whites. I believe they do this because they feel safe resting with similarly colored animals.

Alpacas Have a Social Hierarchy

The alpaca herd is similar to any community, group, or gang. They have a leader and they have followers. The leader, or alpha, tends to watch over the herd and will influence the behavior of the other alpacas in the herd.

For example, super smart Ariana would always watch over the herd, monitor everything outside of their paddock area, and keep a keen eye on what the alpacas and humans were doing inside the paddock. If she sensed any danger she would start to hum. And if she saw danger outside of the paddock, she would scream an "alarm call" so the rest of the herd was aware of the issue.

What I always loved about Ariana was her ability to tattle on my husband. If she was concerned about something, she would make sure I was aware of it too. If I walked into the barn and she was annoyed with changes my husband was making,

Ariana would run up to me and hum in a certain way to get my attention. Then she'd turn to look at my husband while humming so I knew he was the issue. She hated when Jason would make changes to the barn's interior, and she would tattle on him in an effort to make him stop.

Alpacas Can and Will Spit

Alpacas will spit, and they use this as a defensive mechanism. Spitting is not something that an alpaca walks around doing for no specific reason, and it is generally always at other alpacas.

Why do alpacas spit? Anger, fear, dominance over food, and a lack of personal space are all reasons we commonly see for spitting. Pregnancy will increase the frequency of spitting, and this is simply because the female is trying to protect her unborn baby.

If you ask about spitting in an alpaca group on Facebook, you'll see lots of responses from people. Many will tell you to carry a squirt bottle to "spit back" at the alpacas. I have never done this, and I never will. On our farm I use a stern "no ma'am," I place my finger up in their face when doing so, and I give a clear look of dissatisfaction. And for me, this works both short- and long-term.

Alpacas Make Lots of Sounds

Alpacas use a variety of sounds to communicate and express themselves, and each one has a purpose. Learn your alpaca sounds and you'll have a window into their minds and a much better understanding of their actions.

- **Humming:** This is the primary method of communication, and it starts at birth. When a new cria is born, the mother and baby hum constantly to each other. It is their way to communicate and stay connected. Alpacas will also hum in an effort to express distress or anger. This is especially true if

their living arrangements are changed, they are moved, or they are separated from their herd. Alpacas may also hum if they are curious, happy, cautious, or sick. By now, you might be wondering how you would know the difference, but you will quickly learn the sound and meaning of each hum if you listen and observe. I know my alpaca hums and what they mean because I've spent time sitting and listening to them.

- **Snorting:** Alpacas will snort when another alpaca is invading their personal space. It's like a warning message to move away. We don't have a lot of alpacas that snort, but the ones that do, snort often.

- **Clucking:** An alpaca will cluck to express concern or to signal friendly behavior. Both Avalon and Stormy were "cluckers" on our farm. They would cluck when they were excited to see humans approach the pasture. Mothers will also cluck for their young crias and use this as a method of communication.

- **Alarm call:** I had never heard this until Zula arrived at our farm. If Zula saw something she interpreted as danger, she would signal the herd of the threat by using a very high-pitched scream. I think most of our county could hear her alarm calls. It's hard to describe this sound, but when you hear it, you will know it is the alarm call.

- **Screaming:** An alpaca scream is extremely loud. It's as if someone set off a siren next to your head. Unlike the alarm call, a scream sounds like it is backed by anger or fear. Alpacas will scream when they are handled incorrectly or when they *believe* they are not being handled properly. Kalista will scream during shearing no matter how gentle the shearing team is with her. Dream Legend would scream anytime you tried to handle her daughter Sherry Ann. The best part is when Dream Legend would scream, she knew to position her head right by my husband's ear so it would have the biggest impact.

- **Orgling:** An orgle is a sound the male makes during breeding. It is not pleasant and something that would send a human female running. The alpaca female doesn't mind. In fact, the sound does just the opposite. It seems to energize her and get her very excited about the breeding process.

Alpacas Can Open Gates, Turn Off Fans, and Turn On Lights

One day my husband and I were having our morning coffee when one of us noticed chaos outside. The alpacas were running all around and no one was in the right paddock area. Somehow everyone was loose, and no one was claiming responsibility.

We immediately assumed our son forgot to close a gate. That would have explained one gate being opened, but not multiple gates. After we moved everyone back to their correct paddocks, we turned on the camera replays to see what happened. To our surprise, little Nellie had gone around and opened every gate. At six months old, she figured out how to open her gate, then used this knowledge to open *all* of the gates. Sweet little Nellie. We couldn't believe it.

A regular session of providing treats can turn into an adventure when Cammi is involved.

Nellie was just prepping us for Cammi, Petiti, Amara, and Luna. These ladies open gates, steal hats and bowls, and turn fans and lights on and off. We had to put tape over power cords and light switches to secure them, as well as extra latches on gates to make it stop.

You may consider this bad behavior, but you have to remember that alpacas are smart, and they get bored. When they are bored, they can find mischief.

Alpacas Take Note of Human Interaction

If you look at our farm photos, you'll clearly see that we spend a lot of time with our alpacas. My husband is with them at least twice a day for chores, and I go out to the barn each day if my schedule allows. We both view this as an important part of alpaca ownership. It is also an important part of teaching alpacas about humans.

This is me sitting with Nibbler when she first came to the farm.
These little sessions were specifically designed to grow trust.

Even though alpacas can tell one human from another, your interactions influence their perception of *all* humans. You can either build up trust or tear it down with each encounter. I choose to build trust and respect, while also encouraging anyone who comes to our farm to do the same.

When we purchase an alpaca and it arrives home to our farm, I am present to welcome it. As the alpaca walks off the trailer, I make sure I greet it verbally, bend low, and make eye contact so it knows I am not a threat and it is safe. Then I start to naturally interact with our existing alpacas so the new arrival can see how they respond to me and sense this is a safe place. Again, my husband thinks I'm a bit foolish, but I know these practices aid in reducing stress upon arrival and speed the alpaca's acclimation to our farm.

Alpacas Want to Build Love and Trust

I strongly believe you must build trust and love before you ever try to work on behavior modification with your alpacas.

When we purchase or birth an alpaca, I take the time to first build trust with them, so they know they are safe with me. Once they trust me, I then attempt to build love. Alpacas cannot love you until they feel comfortable around you; once you create that level of security, you can create a bond where love grows. Love will grow quickly if you let it.

Once my alpacas trust and love me, I kick into training mode and start working on behavior modification. When an alpaca trusts and loves you, it will want to please you. When it wants to please you, it will voluntarily change its behavior to what it knows is desired and acceptable.

Alpacas Can Be Introverted or Extroverted

Just like their human counterparts, alpacas can be an introvert or an extrovert. In fact, I tend to find it rare that one is middle of the road.

As soon as I head toward the paddock, my favorite girls will run my way to investigate, say hello, and get some love. Nibbler, Sherry Ann, and Jalapena are core members of my posse. They all want and need attention. If I stay in the paddock, all three girls stay with me. If someone is with me, the three girls require the visitors to interact with them as well. All three of these ladies are clear extroverts.

Amara, Kolette, and Monica Ann will also be eager to come over, but these girls are a quick hello or kiss, and then they go back to their business. Those ladies are still extroverts, but they are not as needy as the ladies in my posse.

Amber Fantasy, Faith, Onyx, Daisy, and Peruvian Amber are all introverts. They are sweet, but they don't crave interaction like the others, and they don't need to be the center of attention. They will interact occasionally, but not nearly as often as the others.

Vicki and her daughter Jenna have been the anomalies in our herd. I classify them as introverts who long to be extroverts. Both ladies can be found in the mix with my

Nibbler is a total extrovert!

posse and really want to be part of it, but their quiet personality is overshadowed by the extroverted alpacas.

I didn't catch this at first with Vicki, and I feel bad it took me months to realize what was happening. But now I know, and I've adapted *my* behavior to accommodate Vicki's emotional needs. I made sure she was part of activities and I gave her personalized attention and kisses, but I also never would try to pet her because I knew that would cross the line with her. I would also make sure I shielded her from the extroverts that overpowered her.

I'm thankful I could quickly see Jenna was a carbon copy of her dam's personality. Just like Vicki, she will run to see me and give me a quick kiss, but she has limits. No petting is allowed, and if I'm not watching, she will quickly get overpowered by the personalities of the extroverts in my posse.

Alpacas Can Have Best Friends

When we first started breeding alpacas, we had two babies called Chase and Sherry Ann. Chase was white and Sherry Ann was white with large fawn spots. Since alpacas tend to hang out with their own color, Chase and Sherry Ann were very bonded as babies. When they turned six months old, it was time to wean the crias, and we had to remove Chase from the girl's pen. Sherry Ann was so upset that her best friend was removed, it broke my heart. She showed visual stress immediately: She whined and hummed, and then she lay at the fence line looking for Chase. She sat there for days.

Sherry Ann eventually replaced Chase with Lilly Grace, but it wasn't quick, and they never had that inseparable relationship she had with Chase. Thankfully, once moved, Chase became very bonded with my son—Hunter became Chase's new best friend. In Chase's eyes, Hunter was no Sherry Ann, but he was a suitable substitute.

Alpacas Can Have Mortal Enemies

Nibbler and Amara are half-sisters. They share the same mother and have similar personalities. I should also note that they have very *strong* personalities.

These two girls truly hate each other, and we often can hear Amara squealing in the house with anger over something her sister did. Let me add that our house is far from the barn. I should not hear Amara, and yet I do.

In Nibbler's defense, her little sister Amara is usually the instigator and the one to blame. I keep thinking this will calm down, but it hasn't. I believe their strong personalities collide and they rub each other the wrong way. Thankfully, Amara is now with our mentors at their farm, where she is expecting her first baby. The time away from each other and Amara's pregnancy should alleviate or reduce the conflict when she returns.

Alpaca Can Have No Sense of Personal Space

Our alpaca herd does not understand they are large animals, which means they have no sense of personal space. They'll bump into you without trying, or they'll crowd around wanting attention and annoy each other because they are all much too big for the space.

They'll get right up in your face, again not realizing their size and that you might want a few inches to breathe fresh air. With our herd, you don't get fresh air. Instead, you get hay breath from their exhales.

Have I oversocialized them as babies and caused this issue? No, not at all! It happens with alpacas that were purchased as adults, which means I could not have imprinted on them as babies. What I did do is build their trust and allow them to feel safe when close to me. This sense of safety tears down their walls of protection, which helps them forget they are livestock. They no longer feel like they need a large open space for protection.

Alpacas Can Have Great Memories

A few years ago, our insurance agent introduced us to a widow named Angela. No matter how hard Angela tried to keep up with life and taking care of her herd, she didn't have enough hours in the day to make it happen. She needed to disperse her herd, and we helped rehome all the alpacas except for eighteen-year-old Amber. We knew no one would willingly take an alpaca that old, so we kept Amber and promised Angela we would love her, care for her, and let her live out her remaining years at our farm.

One day, Amber's original owner arrived at our farm. Mary Jane was the retired owner of Big Willows Alpacas in a town south of us. My husband knew exactly who she was and was excited to meet her. After all, Mary Jane's farm was *the* alpaca farm in northern Michigan years and years ago. Jason told Mary Jane that he thought we had one of her original alpacas, and she immediately asked to see Amber.

As Mary Jane walked into the barn, she called for Amber. Amber jumped up and rushed over to her. Normally, Amber ran to no one, because she was the queen of our lady barn and only did what she personally wanted to do. Mary Jane was a different story. Even though she had not owned Amber for well over a decade, Amber still knew her and was eager to go see her.

My husband was amazed at the connection our grannie alpaca had with Mary Jane. I viewed it as a sign of Amber's love, respect, and appreciation for a life well spent on Mary Jane's farm. And it shows alpacas have an amazing memory.

Alpacas Can Maintain a Serious Bond with Their Favorite Humans

I have strong bonds with my alpacas, and while I take pride in this fact, I know I'm not the only human to achieve this. The story of Mary Jane and Amber is clearly an illustration of bonding, but there are so many more.

We purchased Leviticus from Kim and Nancy. We've had him for a few years now, and while I can tell Leviticus loves and respects me, I will never be Nancy. There is no doubt his true love is and always will be Nancy.

Nancy can arrive at our farm and call Leviticus's name from yards and yards away. That boy will perk up, look for her, and be super excited to see her. When she reaches his paddock and greets him, you can see a true connection between them. This big old alpha male will lay his head against her and snuggle in. His love for her is evident in every movement he makes.

I will never have this connection with Leviticus, and that is okay. Nancy birthed and raised Leviticus, so there is no way anyone else can replace her in his heart. I appreciate this and play my role as Nancy's runner-up.

Alpacas Can Be Great Travelers

When Dolly was pregnant with Ollie, she developed a jaw infection. We consulted with a specialist from Ohio State University, who informed us surgery was the only way to save her. OSU also told us to wait until she delivered her baby and then

I'm no Nancy, but Leviticus does love me.

immediately bring her down for treatment. We were worried about her cria, so they told us to bring him along.

OSU is a fabulous hospital for alpacas, so we didn't question their advice. When the day came to head to Ohio, we were having excessive heat in the Midwest. We didn't have air-conditioning in the livestock trailer, and there was no way Dolly and

Dolly and Ollie on their way home from OSU's vet hospital.

Ollie could ride in 96-degree heat. Our only choice was to put them in the back of my SUV.

I am a person who needs things to be clean and neat. The idea of putting two alpacas in my expensive SUV didn't sit well with me. However, my love for my SUV was overshadowed by my love for Dolly and her son Ollie, so we pushed on and hoped they would do well.

It was a seven-hour drive each way. We didn't have any accidents (the potty kind), and when Ollie was hungry we'd just pull over so Dolly could stand and feed him. Intuition told me Dolly would respond to Ollie's needs if we stopped, and Dolly didn't disappoint.

Dolly's surgery was a success, and after a week at OSU we were allowed to go get her and Ollie. Another farm was headed to OSU and offered to pick them up for us. Much to my husband's disappointment, I refused. A fifteen-hour round-trip was worth it to me. I had dropped Dolly off and I wanted Dolly to see my face when it was time to go home. When Dolly saw me at OSU, she started to sprint toward me. She knew I was her owner and that it was time to go home. Her facial reaction and quick movements taught me so much, and it reinforced our bond and increased the level of trust she had for me.

Berserk Male Syndrome

Before closing out our discussion of alpaca behavior, I need to mention Berserk Male Syndrome. This is also known as "Novice Handler Syndrome" or "Berserk Alpaca Syndrome." It is an unfortunate situation that occurs when a young alpaca is mishandled by humans. Many times, it is a male alpaca, although it can happen with females as well.

If a human roughhouses with a young male, the alpaca can interpret this as play and begin to interact with the human as if they are an equal or peer. If the alpaca believes you are equal to them, it could attempt to dominate you or just play really rough, like it would with a fellow alpaca. This leads to the alpaca trying to mount or wrestle with the human, which in turn leads to broken bones and other physical damage for the human.

Make no mistake: Berserk Male Syndrome is caused by humans. Alpacas are not aggressive animals, and their default behavior is not aggressive or dominating. This change in personality occurs when the human has altered their behavior in a negative way, and it produces long-term issues that are usually not reversible.

Our first year of breeding, we had two young males named Teddy and Levi. Teddy was an instigator and Levi was a lovebug. We would interact with them, creating a bond with them, but closely watched for signs of aggressive play or a desire for dominance. I had two incidents where Teddy and Levi went from being loving to wanting to dominate me in play. I quickly stopped the behavior and made sure I did not encourage it moving forward. I don't think either alpaca was being overly aggressive, but there is a fine line between play and male dominance. Both Teddy and Levi grew up without behavioral issues.

As the human, it is your job to maintain the appropriate boundaries so young alpacas do not confuse you or anyone else with a fellow alpaca.

Tips for Making Friends with Even the Most Difficult Alpaca

Bringing your new herd home is fun and exciting! I felt pretty giddy when our first set arrived, and I still feel this way when any new ones arrive. I know you'll feel the same, and I'd like to give you some tips to make this experience positive for all creatures involved.

Remember that every interaction matters. From the moment your alpacas arrive, remember this is an opportunity to build trust or tear it away. This is key. You cannot let your excitement or frustration take over.

Get your big human body on the same level as the alpaca so you are not intimidating. For the ladies, I tend to sit right on the ground but not overly close. This makes me more approachable and a lot less scary. For breeder males, I use more caution but still make sure I bend down to their level when interacting.

Look the alpaca in the eye and talk to them with genuine affection. Some behavior experts advise against eye contact, but for me it is critical. I believe this lets the alpaca know they are seen and they matter, and it helps them know I'm taking the time to interact with just them. Since alpacas are emotionally intuitive, I believe they sense my desire to bond with them, so they never view my actions as a threat.

Say their name as you speak to them. As I walk through the barn or the paddock, I greet everyone by their name. This could be their full name or an abbreviation of it. My girl Nibbler is only ever called Nibbler if she is in trouble. Otherwise, she is Nibs or Nibby. She knows these are her "pet names" and that they are said with love and affection.

In your initial interacting, keep those human arms in and close to your body. This is one of the most important tips. By keeping your arms close to your body, you are reducing movement and keeping yourself small. This shows the alpaca you don't have intentions of grabbing or holding them, which helps show them you are safe.

Just talk to them. Make sure your voice is heard. I've learned my alpacas like me talking and they tolerate me singing. I ask them how their day is, I ask the pregnant girls how they are feeling, and I tell them I love them. I know it sounds ludicrous, but these interactions work for me, and it helps me build a bond with my herd. When we are doing medical treatments and I talk to them, I can see the sound of my voice helps them relax.

Use food to build trust. We will hand-feed their daily food pellets as treats, and many of our alpacas love carrots or apples cut in small strips (like a french fry). This allows close interaction and reinforces the idea that having a human close is a safe activity.

Let them get close. When an alpaca finally starts to voluntarily approach you, keep your hands in, stay still, maintain eye contact, and let them bop your nose, smell your hair and skin, or nibble on your face. The nose bop takes getting used to, but this is an alpaca kiss, and it is a sign of affection. Most of our herd kisses me. Kalista snorts every time she kisses me. Some alpacas come in way too fast for kisses and about knock my nose off, and others give me a big bunch of hay breath. When you're building trust, you take all of it and accept them, even with all their goofiness. I also thank them for the kisses, so they know I appreciate the love.

Watch your surroundings. You want to make sure the alpaca isn't in a corner or backed into a building, fence, or gate when you approach them. This location will make them feel trapped. They need to know they have an exit plan if needed. The reality is they likely won't leave, but they still want to know an escape route exists.

If your alpaca is bad, tell them so. To correct misbehavior, I use a "mom voice," put my finger up, and scold them using their name. I use a consistent phrase along with their name, such as "no ma'am," so they know that behavior isn't acceptable. This trick is super effective, and as I've mentioned, some of the alpacas just have to see my facial expression change to know they've crossed the line. Alpacas are smart and they learn quickly.

How long does it take to woo an alpaca? It highly depends on the alpaca's age, prior experience with humans, and overall personality. Just like humans, there are introverts and extroverts, some have experienced trauma, and some are just aloof or timid. The important thing is no matter what the situation, you can build trust if you work at it.

CHAPTER 8

Taking Care of Medical Issues and Herd Health

Herd health is an important topic for new alpaca owners. This subject encompasses not just ongoing care, but also the assessment of health and treatment when issues arise.

In this chapter, I will offer some tips for monitoring your herd for issues, discuss basic ongoing care, and review some common parasites you need to be aware of.

Monitoring Your Herd's Health

Many farms and ranches will have a monthly "herd health day" to review and assess their alpacas. For some farms, this is really the only option for monitoring health. It is systematic and structured, and everyone is evaluated within a specific amount of time.

Other farms, including ours, take a different approach and will do daily reviews of appearance, personality, and activity to assess health. We then do a little deeper review each month when we give each alpaca their preventative ivermectin injection.

Regardless of the camp you're in, it is important to be proactive in health care and evaluation. Since alpacas are prey animals, they default to hiding illness. This means an alpaca can go from seemingly healthy to very sick quickly. We've had it happen, and it is heartbreaking.

The important takeaway is you have the ability to reduce the risk of severe illness. You just need to remember that alpacas are unique, and as their caregivers you need to actively monitor and check for subtle signs and issues.

My husband is with our alpacas twice a day for chores, so during this time he monitors them and their activity for signs of illness. If anything is off, I'm called out to the barn to do a walk-through to observe and evaluate. Because I am a people-watcher,

I pick up on small nuances in behavior and appearance, so I can see what my husband might miss.

If I walk through the paddocks and have the slightest concern with any alpaca, or if my husband calls me out to review a particular alpaca, I will stop everything I'm doing to quietly observe. I need to compare what I see and hear to the alpaca's standard behavior. I first review the alpaca's face for any signs of illness or fatigue. Then I attempt to interact with the alpaca to see if its behavior is different from the norm. Last, I stay quiet and watch this alpaca and the other members of the herd around them. If I suspect something is wrong, we do a more thorough review and evaluation.

Here's the important part that I really want you to remember: While I'm fairly spot on with catching illness or problems early, I can do so because I know my herd deeply. For example:

- I know Sherry Ann has a gurgle, and if she is annoyed it gets louder. This is not a sign of illness. That said, I know we have an issue if she does not push her way into the herd to monopolize my attention.

- I know Kalista always snorts when she kisses me, and this doesn't mean she is congested or angry. If she doesn't want to say hello and come in for a kiss, I might have an issue on my hands.

- I know Adel (and all her children) sleeps like a rock, and this doesn't mean she is lethargic. For Adel, I just need to get down on her level to see if she is annoyed with my presence. If she is, all is well. If she doesn't care, we might have an issue.

- I know Auburn will always run over to kiss me, and if she doesn't, it's a sign of a problem. If she fails to proactively interact, this normally indicates her fiber needs to be trimmed near her eyes. When she cannot see well, she gets anxious, and anxious girls won't run around or freely give kisses. If we were to trim around her eyes and she still fails to interact, I know we might have a larger problem.

- I know none of my girls will bother to move when I walk through the barn, and this is never a sign of lethargic behavior or illness. This is just them being comfortable with my presence. If my husband walks through the barn with me and someone doesn't move, I know we have an issue. They never sit still when we are both in the barn because they know that is a signal that someone needs medical attention.

- I know if I bring carrots out to the barn and all hell doesn't break loose, we have a major issue and everyone is getting evaluated. Carrots and apple slices create a lot of excitement. If chaos doesn't happen, something is wrong, and we need to investigate.

Because I know the ladies and all their individual "tells," I know when something is wrong. It's this level of intimacy that helps us keep our herd healthy. And the interesting thing is I'm not alone in this. There are many alpaca owners who are just like me. I think the reason for this is alpacas are lovable creatures with strong personalities. Many owners love to spend time with them, so by default they know their herd's personalities and develop superpowers for noticing issues quickly.

Adel's son Glenn is napping like a regular alpaca. This is rare for Adel or any of her children.

Common Signs of Illness

When you first bring your new herd home, you won't know enough about each alpaca to identify illness quickly. While you get to know them, you should watch for both healthy behavior and signs of illness.

Signs of a Healthy Alpaca

- Normal eating and drinking habits.

- Alertness and engagement with you and the other members of the herd.

- Clear and alert eyes without discharge or signs of infection.

- Ease of movement, with no limping, stiffness of joints, or hampered movement due to overgrown toenails.

- Normal respiration sounds, with no excessive panting or labored breathing.

- Healthy poop pellets that are not runny or clumped into large balls.

Signs of a Sick Alpaca

- Not eating and drinking.

- Lethargic and not wanting to get up or move.

- Separation from the herd. Alpacas have a classic case of FOMO (fear of missing out), so they rarely spend time alone. Unusual separation from the herd is a classic sign of illness.

- Empty eyes, goopy clumps in eyes, or discharge.

- Limping, stiffness of joints, or difficulty moving.

- Strange respiration sounds, panting, or labored breathing.
- Clumped poop or diarrhea.

Performing a Hands-on Evaluation

Should you notice anything amiss, you'll need to do a hands-on evaluation that could include one or more of the following items:

- Review the fiber coverage around the alpaca's eyes to see if this is causing a decline in vision, which can—and generally will—impact activity and behavior.

- Try to get the alpaca up and moving if it is lethargic.

- Check the alpaca's temperature to see if it is above or below normal levels. An alpaca's normal temperature range is between 99.5°F and 101.5°F for adults.

- Check the body score of the alpaca to assess possible weight loss (see page 116).

- Weigh the alpaca on a livestock scale to see if there have been any changes from the last recording.

- Test for anemia by checking the FAMACHA score (see page 114) and comparing the inner eyelid color to the official color card.

- Inspect the jaw, teeth, and mouth for any lumps, broken teeth, or other visual issues. Also, review any adult males for fighting teeth that may have erupted or need trimming. Fighting teeth are sharp teeth located in the back upper jaw. They normally arrive between the ages of two and three.

- Review the poop within the paddock or barn for any signs of parasites or illness.

When in doubt, call your mentor or vet to discuss your observations and review next steps.

Tips to Prevent Issues

- Provide quality hay that is free of sharp weeds or stems.

- Provide free choice minerals (phosphorus, calcium, selenium, copper, and zinc) so the alpacas can self-moderate vitamin intake.

- Perform monthly herd health checks or daily observations.

- Clean up manure and remove it daily.

- Perform annual shearing that takes place just before the summer heat arrives.

- Provide CD&T injections at the annual shearing to protect against diseases caused by *Clostridium perfringens* type C and type D, and *Clostridium tetani*.

- Trim toes as needed, which may be every three to four months. Some alpacas require it much more often than others.

- Monitor the length of eye fiber and trim it when needed.

- Perform periodic fecal tests. This can be done at home (equipment permitting), by your vet, or sent off to a lab.

- Keep alpacas separate from other livestock to reduce parasite infections and bodily harm.

- Provide fans in summer to help avoid heat stress in warmer climates.

- Provide coats for elderly alpacas in winter.

- Only halter alpacas when needed, and never leave a halter on long-term. Extended halter use will cause unnecessary stress, increases the risk of strangulation, and can cause issues with fiber growth around the face.

What Alpaca Owners Need to Know About Parasites

When we first started our alpaca farm, we knew *nothing* about parasites. We were city people, and being able to identify parasites in alpaca poop wasn't on the list of skills we had acquired in corporate

Yvette testing out her healed leg after an accident and weeks of wearing a cast.

life. Thankfully, we were very fortunate to have Rebecca Fowler from Oasis Acres in southern Michigan take my husband under her wing and teach him the ins and outs of parasite management. She not only taught him what to look for in symptoms, she also taught him how to perform fecal testing on his own to evaluate which parasites were present. This not only helped keep our herd healthy but helped save money on vet bills, too.

Since you won't have a Rebecca Fowler available, I'd like to provide some basic information to get you started.

Alpacas are generally hardy animals, but like any livestock they can fall victim to a range of parasites that can impact their health. Following is an overview of the most common parasites affecting alpacas. Please note this is *not* an exhaustive list or a replacement for veterinary care. It is a guide to help you understand what you need to be aware of in managing your alpaca's health.

DEWORMING SHOULDN'T BE PREVENTATIVE

Outside of monthly injections for m-worm, never proactively deworm or treat your alpacas without knowing what parasite is present and which alpaca the parasite has impacted. Proactive deworming leads to resistance to medications, which in turn can cause issues with all your alpacas.

Meningeal Worm

The meningeal worm (also called m-worm) is a parasite that primarily affects alpacas and other camelids. M-worm starts with white-tailed deer and is transferred to alpacas via snails or slugs. The larvae then migrate through the alpaca's nervous system, lead-ing to neurological symptoms such as weakness in the rear legs, lack of coordination, stiffness, and paralysis.

In Michigan we have a large population of white-tailed deer, so we proactively give monthly ivermectin injections to protect our herd against this deadly disease. Consult with your local veterinarian to see if this is something you need to do in your area.

M-worm is fast and deadly. Treatment must be aggressive and swift, so you should immediately consult a veterinarian if your alpaca shows signs of infection.

Tapeworm

Tapeworms are intestinal parasites that can infect alpacas, especially if they ingest contaminated pasture or feed. The adult tapeworms release eggs that are eaten by mites, which then become ingested by an alpaca. Tapeworm infections are often asymptomatic—you'll find a worm in a dung pile, but the alpaca will have no other symptoms. On our farm, a tapeworm will upset me way more than it will the alpacas.

Barber Pole Worm

Barber pole worm is a parasite that can cause anemia and severe weight loss in alpacas. It thrives in warm, moist environments and is commonly transmitted via a contaminated pasture.

A FAMACHA test (more on this below) is a good way to help assess if this parasite is impacting the health of your herd. Like m-worm, barber pole worm needs to be treated swiftly and aggressively.

Small Coccidia

Coccidia is a parasite that lives in the intestines of alpacas. While many species of coccidia are harmless, some can cause coccidiosis, leading to gastrointestinal upset and diarrhea.

Daily manure cleanup and dry living spaces will help reduce this parasite. Avoid overcrowding and ensure alpacas have enough space to avoid stress, which can compromise their immune systems.

E. mac

E. mac (*Eimeria macusaniensis*) is a parasite that is found in both adults and crias. Lethargy and weight loss are common signs of this condition. If left untreated, *E. mac* can cause death, so quick identification and treatment is necessary.

Liver Fluke

Liver flukes are parasitic flatworms that can infest the liver of alpacas, particularly if they graze on wet, marshy pastures where fluke larvae are commonly found. Liver fluke is primarily found in the western United States, which means our farm has never encountered this parasite. Consult with your vet to see if this is something that is common in your area.

Mites

There are a variety of mites that infect alpacas. Mites are transferred via direct contact, bedding, and dust piles. They burrow into the skin, causing itching and irritation. Symptoms include scratching, hair loss, and thickened skin. Mite reactions will vary in degree from alpaca to alpaca. Some alpacas seem greatly impacted, while others seem to have no symptoms at all.

Treating mites requires knowing what kind of mite you are dealing with. Therefore, it is best to get an experienced vet involved with diagnosis so you know you are using the proper protocol.

Flies

Flies can be a nuisance for alpacas, causing irritation and sometimes leading to secondary infections. We've found the best treatment is prevention in the form of daily poop removal, fly traps, and fans. We also use Fly Predators® from Spalding Labs throughout the spring and summer months. These little bugs arrive in the mail, and we spread them around our property so they can provide a natural method of proactive fly control.

THE 80/20 RULE

Most experienced alpaca vets state your herd will have 80 percent of the parasite issues in only 20 percent of the alpacas. Once you find the 20 percent of the alpacas that are impacted, you'll want to keep a very close eye on them.

Two Tests for Keeping Your Herd Healthy

There are two tests you'll use repeatedly in herd health. These are FAMACHA scoring and body scoring. Neither of these are difficult to learn, but they are very important for maintaining a healthy herd.

FAMACHA Scoring

FAMACHA (pronounced "FAM-uh-sha") is a simple method of evaluating the color of the mucous membranes (inner lining of the eyelids) to determine anemia. This scoring system was first developed by Dr. Faffa Malan for use in goats and sheep; however, it was later adopted by alpaca owners.

The test is fairly easy and quick if performed outside (if possible) and in sufficient light. To perform the test, you will compare the alpaca's lower eyelid to the official FAMACHA score card.

The FAMACHA scale ranges from one to five, with each score indicating a different level of anemia:

- Score one (red) suggests a healthy alpaca that is not anemic.

- Score two (pinkish red) suggests normal health, but you should use caution and monitor the alpaca's health.

- Score three (pink) suggests mild anemia, which could indicate the alpaca is showing early signs of blood loss due to parasitic infection. At this point we would perform a fecal test to validate.

Jason is checking Ryan's FAMACHA score using the official score card.

- Score four (pinkish white) suggests moderate anemia. We would test a fecal sample for parasites, deworm, and then monitor the alpaca closely.

- Score five (white) suggests severe anemia. We would test a fecal sample for parasites, deworm, and contact our vet for an on-site medical review.

While FAMACHA scoring is a useful tool for detecting anemia due to parasitic infestations, it is not a stand-alone activity. If an alpaca in our herd were to test high, my husband would immediately drop everything he had planned for the day and launch into protection mode for our herd. This would include:

- Testing fecal samples for this specific alpaca.

- Checking FAMACHA scores on other alpacas within our herd.

- Performing fecal tests for all paddock areas to make sure we do not have a growing barber pole worm crisis.

- Contacting our vet with the results and requesting an on-site visit if conditions warranted.

If you are new to FAMACHA scoring, I would recommend you speak to your local vet to discuss training, obtain a score card, and create an action plan for testing, evaluating, and addressing concerns.

Body Condition Scoring

Body condition scoring (BCS) is a fast, hands-on test to assess the alpaca's overall health. It provides a physical observation that is not skewed by fiber length or density. The scoring system ranges from one to five, with a score of one indicating emaciation and a score of five indicating obesity. The ideal BCS for alpacas is typically three, which indicates a healthy, well-balanced body condition with adequate muscle mass and fat reserves.

- A score of one would indicate an alpaca is severely underweight, with prominent ribs, spine, and hip bones. The alpaca is at risk for malnutrition and requires immediate intervention.

Alpaca Body Condition Scoring Guide

Mazuri

Description	Score	Side View*	Top View*	Front*	Back*
THIN: Deep depression on either side of dorsal vertebral processes. Sunken-in flank, with bony "shelf" just above it. Sharp shoulder, withers, and hip bones; prominent ribs. Prominent V-shape to keel, and sharp inverted V between rear legs.	1.0 / 1.5 / 2.0				
APPROPRIATE: Slight cover over bony structures; ribs and spine palpated with slight pressure. Body contours visible but smooth in appearance. Healthy muscle mass. Slight to moderate fat in keel and between rear legs.	2.5 / 3.0 / 3.5				
OVERWEIGHT: Bone structure palpable only with moderate to firm pressure, or not palpable. Little to no body contour visible; rounded appearance. Fat pads visible over keel, between rear legs, and possibly around tailhead.	4.0 / 4.5 / 5.0				

*Appearance will change based on fiber length and coverage, as well as degree of pregnancy. Manual evaluation is important.

Adapted from: PennState Extension, Body Condition Scoring of Llamas and Alpacas, Robert J. Van Saun, DVM, MS, PhDr | Australian Alpaca Association 2008. Alpaca Note #4: Body Condition Scoring (BCS) of Alpacas. | Edmonson et al, JDS 1989,72:68 and Russel, A. Body condition scoring sheep, Sheep and Goat Practice 1991.

Mazuri's body scoring guide, which was adapted from a version available from PennState Extension. IMAGE COURTESY OF MAZURI.

- A score of two would indicate the alpaca is thin. The ribs and spine are still visible, though not as prominently as in a score of one. Some muscle definition is present, but there is little to no fat covering. This alpaca would benefit from increased food intake or a change in diet.

- A score of three would indicate the alpaca has a healthy, well-rounded appearance. Ribs may be felt but they are not easily visible, and there is a light layer of fat covering the ribs, spine, and hips.

- A score of four would indicate the alpaca is overweight. The alpaca is visibly fat throughout the ribs, spine, and hips. The alpaca may be at risk for obesity-related health issues, such as metabolic disorders.

- A score of five would indicate the alpaca is obese. The alpaca is very overweight, with significant fat covering the ribs, spine, and hips. This alpaca is at high risk for health problems such as joint strain, metabolic issues, or cardiovascular strain.

I confess, on our farm our alpacas tend to be in the range of three to five in body condition scoring, and I am to blame for any fours and fives. We are not benchmark with this scoring and we know it. The issue is hard to modify because my husband and I are so worried about having anyone skinny or sickly, we overfeed hay.

And sadly, the impact of a high BCS isn't isolated to just the issues listed above. Overweight alpacas will have "fiber blow-out," which is a term to describe a rapid degrade in fiber quality. We have experienced this on our farm, and we know the humans are to blame.

For the health of your alpacas, don't be like the Gill family. Manage food intake to help keep your alpacas at a healthy body score of three!

CHAPTER 9

Diving into Fiber Characteristics and Color Genotyping

When we first started visiting alpaca farms and shows, I kept hearing references to histograms and EPDs. I was pretty darn confused about these terms, because they were generally used individually and not in relationship to each other. I knew they were in reference to an alpaca's fiber quality, but I didn't understand the nuances of the results.

I'm going to walk through both in this chapter, so you have a better understanding of these two techniques and how they pertain to an alpaca's fiber quality. This information will be a great precursor for our next chapter on breeding alpacas.

Histogram vs. EPD

In the most basic sense, these two terms are broken down into these simple definitions:

Alpaca histograms are related to fiber quality at a given point in time. This helps an alpaca owner know what the fiber can be used for once harvested. It's a snapshot of the health of the alpaca, and the data helps alpaca breeders make better decisions.

Alpaca EPD data provides information about the potential quality of an off-spring's fiber. This helps an alpaca breeder know the genetic worth of an alpaca, as well as which alpacas should be bred and to whom they should be bred.

Stated another way, a histogram is all about the here and now, while an EPD provides additional information about what an alpaca can produce. Let's dig deeper into both.

What Is a Histogram?

A histogram is a measurement of quality for an alpaca's fiber at the time the sample was taken. In the United States, we shear annually, so this time frame would be a year's growth. In other countries, like Peru, they shear every eighteen months, so their time frame would be different.

Many alpaca farms will take a small sampling of the fiber during the shearing process and mail it to a lab for testing. This testing will cover key characteristics of the fiber, such as length, micron count, staple length, and curvature. A histogram will also include metrics for uniformity, crimp, strength, and density. I know that all seems like a lot of technical jargon, but stay with me. I'll break these down into what they mean and how to use this data.

Mean Fiber Diameter (Microns)

The mean fiber diameter, or MFD, represents the thickness of a strand of fiber, measured in microns. A lower micron number represents a finer fleece and higher-quality fiber. In the United States, we typically see alpacas born with a micron count between fifteen and twenty-five.

When producing alpaca products, we will use this micron count to label the fleece into grades (like one, two, or three) or names like royal, baby, or superfine. These labels help product producers know what they can make from the fiber. This will be discussed further in Chapter 13.

This micron count will progressively move higher as the alpaca ages, if they become sick, or if they overeat. Some lineages have a very slow increase in MFD, while others will rise rapidly.

Standard Deviation (Microns)

Standard deviation (SD) is the amount of variation occurring in a fiber sampling. The lower the number, the more uniform the fiber and the better the quality of fiber available for producing products. We like to see this number in the threes or low fours, so that when the raw fur is all blended together it creates a soft and consistent feel.

The theory is an alpaca with a low standard deviation will produce softer products. And the really cool thing is that a fiber blanket with a higher micron count and lower

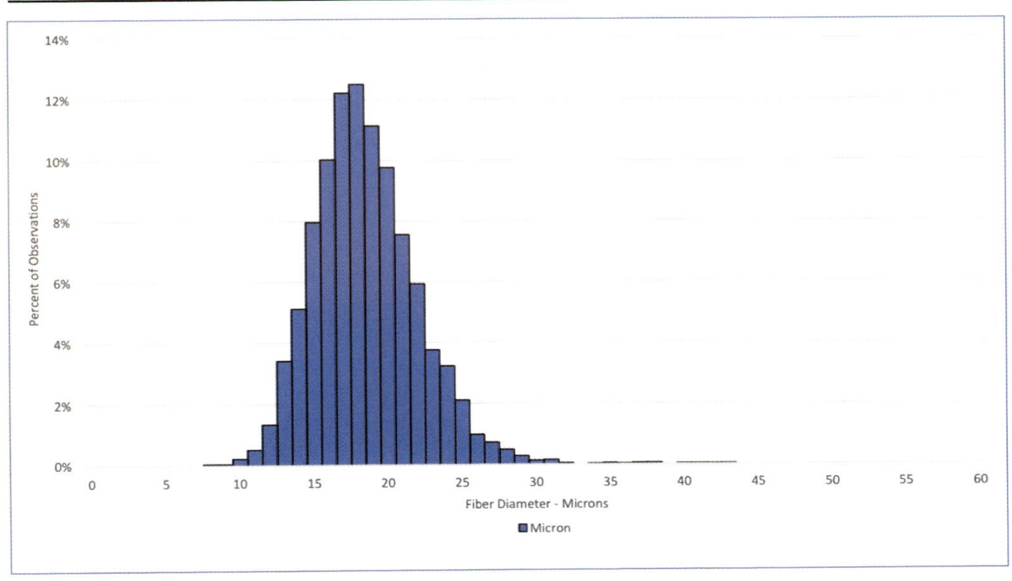

Example histogram report for Snowmass Royal Favor.

SD. can actually feel softer than a fiber blanket with a lower micron count and higher SD. That point can be hard to digest, and I struggled with it for over a year. Our mentor Nancy tried to explain this to me while evaluating fiber blankets in our kitchen and I just wasn't "getting it." Once the raw blanket was made into yarn and I could feel how soft the yarn was, I literally felt what she was trying to teach me.

The takeaway is you want a low micron count, but you must balance this with a low standard deviation. A higher SD is a sign of inconsistencies in the fiber and can significantly reduce the quality in a low micron fiber blanket.

Coefficient of Variation (%)

This data point is the SD as a percentage. A coefficient of variation (CV) under twenty is very uniform, while a CV over thirty is not. We want to target the teens for this number.

Spin Fineness (Microns)

Spin fineness provides an estimate of the performance of the fiber when it is spun into yarn. This measurement combines the MFD and the measured CV. I personally worry much less about this item and focus more on the MFD and the SD.

Fibers Greater Than Thirty Microns (%)

The percentage of the fibers measured over thirty microns is a really helpful data point for both breeding and producing products. In looking at this percentage, you would want to see the lowest number possible.

In younger alpacas with a low MFD, you would expect to see a low figure here, between zero and five. In older alpacas with higher micron counts, you might see this number jump up closer to twenty.

It may feel like this value is a duplication of the data points we've already looked at. It is not, but they all do closely relate to each other and work together to create a high-quality fiber blanket that is perfect for producing luxury garments.

Comfort Factor (%)

In technical terms, comfort factor is the percentage of fiber over thirty microns subtracted from one hundred. Similar to MFD and SD, this helps alpaca owners determine which fiber will make the most desirable yarn and products. The higher the comfort factor, the better the fiber, and the softer it will feel when made into clothing.

Mean Staple Length (mm)

Mean staple length is a measurement of the fiber length. This is a critical measurement, because short fiber (less than three inches for Huacaya alpacas) cannot be used for producing yarn and many other products via the worsted process.

When reviewing this data, you need to convert the millimeter value into inches to understand it. An MSL of 80 mm would equal 3.15 inches. This Huacaya fiber would be spinnable into yarn, but it may not stay this way for long. The older the alpaca gets, the shorter their fiber becomes, so you need to remember this number will degrade over time.

Length Standard Deviation (mm)

This is the SD of staple length—in other words, it is a measurement of the fiber's consistency of length.

Length Coefficient of Variation (%)

The length CV is defined as the ratio of the length SD to the mean of fiber length.

Mean Curvature (deg/mm)

Fiber curvature measures crimp, or the number of waves (zigzags) per inch within the fiber. The greater the number of degrees per millimeter, the more waves you'll see in the fiber.

Finer crimp provides greater resistance to compression and better memory. This is important for the textile industry and producing end-use products at scale. A mean curvature of fifty or above would indicate an alpaca has solid crimp, while a mean curvature of seventy or above would indicate exceptional crimp.

SD Curvature (deg/mm)

This is the SD of curvature and is a measurement of the fiber's consistency of curvature.

Medullated Fibers (%)

Medullation is a fiber trait that is evaluated in light-colored alpacas (white or fawn) and helps quantify the hollow space within the fiber.

Because this metric is often correlated with coarse hair, it can be confused with the more common term of guard hair. While there is a correlation between medullated fibers and coarseness, they are not the same.

Medullation can make the raw fiber firmer, and it can impact the dying process. Therefore, you would want to see a lower number here. High-quality alpacas that are white can many times have a medullation percentage of zero.

What Is an EPD?

EPD stands for Expected Progeny Differences. The use of EPD data was introduced decades ago by Dr. Charles Henderson at Cornell University. Many livestock industries have adopted this data set and regularly use EPD information to make sound breeding decisions.

As I stated earlier, an EPD provides information on what the alpaca might pass on to their offspring. Due to this, it can be a great piece of information for purchasing seed stock or when matching up breeding pairs.

In the alpaca industry, EPDs are used to calculate nine fiber traits and one trait related to birth weight. The current ten traits alpaca EPDs evaluate include

1. Average fiber diameter (AFD)

2. Standard deviation of fiber diameter (SDAFD)

3. Spin fineness (SF)

4. Percentage of fibers greater than 30 microns (%F>30)

Trait	Value	Acc	Rank	Rank%	
AFD	-2.204	0.57	1,026 of 56,478	1.82%	Top 5%
SDAFD	-0.617	0.57	1,121 of 56,478	1.99%	Top 5%
SF	-2.373	0.64	707 of 56,478	1.26%	Top 5%
%F>30	-5.92	0.643	4,979 of 56,478	8.82%	Top 10%
MC	7.773	0.554	110 of 56,478	0.2%	Top 1%
SDMC	3.218	0.589			
%M	-3.894	0.191	3,645 of 56,478	6.46%	Top 10%
MSL	-5.724	0.482	55,570 of 56,478	98.4%	
FW	0.444	0.357	10,371 of 56,478	18.37%	Top 25%
BW	-0.217	0.105			

Snowmass Royal Favor's EPD results

5. Mean curvature (MC)

6. Standard deviation of mean curvature (SDMC)

7. Percent medullated fibers (%M)

8. Mean staple length (MSL)

9. Fleece weight (FW)

10. Birth weight (BW)

Each year the Alpaca Owners Association (AOA) collects fiber data from participating farms. This data is prepared by an independent third-party company using advanced equipment to evaluate fiber characteristics. While this data is generated on individual alpacas, it is also combined to provide metrics on the industry as a whole.

The more accurate data we have on EPDs, the better decisions we can make as an industry and/or alpaca breeder. Stated another way, EPDs help the alpaca industry skip over the trial and error associated with random breeding decisions.

EPDs also help alpaca breeders match sires and dams based on an alpaca's genotype (genetics). This helps us avoid poor breeding decisions based solely on phenotype (appearance).

On our farm, I always pay close attention to both phenotype and personality, while Jason studies the genotype. He focuses on the data, while I look for pleasing facial features, unique coloring, physical characteristics, and personality. At the end of the day, EPDs help my husband keep me from making poor emotional decisions.

Pros and Cons of the EPD Program

You will hear and read mixed reviews of the EPD program. This is because EPDs do have a lot of pros and cons associated with them and their usage.

Some positive aspects of using EPDs:

- They are very helpful in understanding fiber traits and how your specific alpaca is ranked against others in the industry.

- EPD calculations consider factors such as weather and how this can impact fiber traits from year to year.

- EPDs provide a value that predicts heritability and looks not just at your alpaca, but also the other alpacas in the entire lineage. This provides a more holistic view of quality and helps highlight when a genetic line is prepotent for a specific trait such as staple length or curvature.

Some negative aspects of using EPDs:

- EPDs are only as good as the fiber sample. If you or your shearing team pull a sample from the wrong location, it will throw off the data for this alpaca and its offspring.

- EPDs are only as good as the pool of participants. Many farms do not use the EPD program due to cost, so the data is subjective. It is only accounting for those alpaca owners who submit samples, and it cannot provide information for any-thing outside the sample pool.

- To have highly accurate EPD results, alpaca owners need to participate year over year. That requires labor, time, and money.

- Accuracy comes into play. You will see EPDs with low levels of accuracy, which means the pool of data for this alpaca and its relatives is limited.

- It takes many months to obtain your EPD results. This is because the lab needs to test all available samples, calculate these results to determine the official EPD results, then submit the results back to the AOA for loading into the official AOA database.

Color Genotyping

A lot of people assume if you mix a white alpaca with a brown alpaca, you'll end up with a fawn alpaca. I've read this so many times on Facebook and it always makes me laugh. Alpacas are not like mixing frosting or colored wax to make crayons. There is a lot more science that goes into the color outcomes of alpaca breeding.

In our first year of breeding alpacas, we were at our farm with friends Dave and Kari. They are fellow alpaca breeders who brought over their herdsire Eddie to mate our female alpaca Vincentia. As we sat and discussed the breeding combination of Eddie and Vin, Dave made a joke that stuck with me. He compared alpaca breeding to a gumball machine. He said you cannot predict color outcomes of alpacas; you can only hope your preferred color pops out.

At the time I laughed because this funny comment felt so true. But the idea and the image it put in my head never left me. While we can do our best to perform research and match up alpacas for the best possible outcomes, alpaca fiber color is still a guessing game. We cannot accurately predict outcomes because we simply lack the data to do so.

Or, at least, we did. All of that changed in 2021.

In 2021, my husband Jason began working on color genotype testing with our friend and mentor Lynn Edens. This collaborative project was created to better understand the genetics behind alpaca fiber color. The testing was designed to look past the phenotype (what you see) and look at the genotype (the genetic base color you don't see). This project became a game changer for the alpaca industry worldwide.

Jason and Lynn took blood samples from both farms and sent them to a Canadian lab for analysis. This lab was providing a color genotyping test that was developed in collaboration with Dr. Kylie Munyard from Curtin University of Western Australia.

Dr. Munyard is a genetics expert who is known worldwide for her extensive research in alpaca fiber color and genetics. She is active in the alpaca industry, and she continues to improve testing and the data we derive from it. More importantly, Dr. Munyard collaborates with alpaca associations throughout the world to obtain phenotype information to compare to her genotype testing results, which in turn help prove her hypothesis and propel the quality of alpacas forward.

As a bystander, it has been extremely gratifying to watch Dr. Munyard interact with alpaca breeders and associations in such a close manner. She needs our real-world breeding results and we desperately need her research.

Why Color Genotype Testing Matters

An alpaca's fiber quality has always been influenced by the color of the fiber. Historically, white alpacas have much higher-quality fiber than black or grey alpacas. Our goal with color genotype testing is to improve the quality of the black and grey alpaca herds by transferring the fiber characteristics from white alpacas into dark alpacas.

By knowing what the base color of an alpaca is, we can better predict color and control outcomes in offspring. This advancement actually helps us focus less on color—because genotyping now solves that mystery—and more on fiber characteristics such as fineness, density, staple length, and uniformity.

As alpaca breeders of color, genotyping provides the following benefits:

- Provides the genetic base color of an individual alpaca.

- Increases the probability and predictability of expected fiber colors in offspring.

- Identifies mutations and helps us reduce undesirable mutations in offspring.

- Reduces the probabilities of suboptimal alpaca combinations.

- Improves the chances of producing classic greys (which are desirable).

- Improves fiber quality of dark-colored offspring.

- Reduces chances of miscarriages in certain breeding combinations like classic greys.

- Increases farm revenue through the production of higher-quality fiber and more predictable color outcomes.

The Science Behind Color Genotype Testing

Genotyping can sound rather scary, but it doesn't have to be. While I'll walk through the basic science and data here, alpaca breeders really only need to learn the differences between the As and Es.

Dr. Munyard's testing is the process of determining differences in the genotype of an alpaca by examining the individual's DNA sequence and comparing it to another alpaca's sequence or a reference sequence. It reveals the alleles (matching genes) an alpaca has inherited from its sire and dam.

Sample Barcode	Animal ID	Registration Number	MC1R + ASIP Genotype	Grey/Non-Grey Status^	MC1R_C90 1T	MC1R_A8 2G	MC1Rdel_22 4-227ACTT	ASIP_325-381del57_b	ASIP_C292 T	ASIP_G35 3A
62521040548	Cotton Creek Silver	36157211	EE aa	Classic Grey	EE	EE	EE	AA	Aa2	Aa3
62521040549	Cotton Creek Xavier	36157228	EE Aa	Classic Grey	EE	EE	EE	AA	Aa2	A
62521040550	Cotton Creek Leonardo	36157235	EE aa	Classic Grey	EE	EE	EE	Aa1	AA	a3
62521040551	Cotton Creek Francis	36157242	ee AA	Non-Grey	e1e1	e2e2	EE	AA	AA	A
62521040552	Cotton Creek St Patrick	36157259	EE aa	Non-Grey	EE	EE	EE	AA	a2a2	A
62521040553	Cotton Creek Luna	36157112	EE aa	Non-Grey	EE	EE	EE	Aa1	Aa2	A
62521040554	Cotton Creek Olivia	36157129	Aa	Non-Grey	EE	NR	EE	Aa1	AA	A
62521040556	Cotton Creek May	36157143	Ee Aa	Non-Grey	Ee1	Ee2	EE	AA	Aa2	A
62521040557	Cotton Creek Sophie	36157136	Ee Aa	Non-Grey	Ee1	Ee2	EE	AA	AA	Aa3
62521040558	Cotton Creek Henry	36157167	EE aa	Non-Grey	EE	EE	EE	AA	Aa2	Aa3

Color genotype testing results for our crias.

I said earlier that breeding for specific alpaca colors doesn't work like mixing two colors of frosting. This is because alpaca fiber color is complex. It includes two main genes responsible for base fiber color: ASIP color alleles (A or a) and MC1R (E or e), which highlight dilution. It also includes an unknown number of genes involved with pattern fiber. Let's break these data points down even further.

MC1R + ASIP

EE aa
EE Aa
EE AA
Ee aa
Ea Aa
ee aa
ee Aa
ee AA

MC1R Genotypes (Dilution)

- **EE:** The animal does not carry a dilution mutation. It will not produce white or beige offspring.

- **Ee:** The animal carries a single dilution mutation. It will not be white or beige itself but can produce white or light offspring.

- **ee:** The animal carries two dilution mutations. It will usually be white or light in color (medium fawn is also possible when combined with some ASIP genotypes), and it will always pass dilution on to its offspring.

ASIP Genotypes

- **AA:** There are no black mutations present. This animal cannot produce black offspring.

- **Aa:** There is one mutated allele for black fleece present, represented by the little "a." The animal will not be black itself, but it can produce black offspring.

- **aa:** There are two black mutations present. The animal may or may not be black itself but will always pass on a black allele to its offspring.

Predicting Color Outcomes

By using the color genotype test results and what we've witnessed from breeding, we can also calculate the most likely color for each genotype. The following chart shows an alpaca's probability of a fiber color based on the genotype test results. The remainder of the percentage (displayed percent less 100) is a shade lighter or darker. In the first example, 83 percent of alpacas with a color genotype of EEAA will have brown fiber and the remaining 17 percent will be a shade lighter or darker.

Color Genotype	Phenotype Color	Probability
EEAA	Brown	83%
EEAa	Brown	90%
EeAA	Fawn	80%
EeAa	Fawn	87%
EEaa	Black	85%
Eeaa	Black	80%
eeaa	Fawn	90%
eeAA	White	97%
eeAa	White	97%

Now let's take what we know and apply it to four different herdsires. Two are dark and two are light. In years past, we would have believed the white herdsires would produce light offspring and the dark herdsires would produce dark offspring. That is not the case, and we now have the data to prove it. More importantly, we now have the data to control how these herdsires produce colors in their offspring.

When bred to the right dam and genotype, a white alpaca like Moonlight Lover can produce black offspring. This means we can breed him to pass on his exceptional fiber traits (like staple length, mean curvature, and uniformity) to a black-colored offspring. This outcome would increase the quality of a black herd and set up future offspring to have these exceptional traits.

To anyone new to alpacas, that last statement might seem insignificant. To a veteran alpaca owner and breeder, it is a massive shift for our industry. Being able to transfer the premier fiber characteristics of a white herd into a dark herd changes the landscape of alpaca breeding and fiber production throughout the world.

The information I've shared here is just a highlight of what the industry is learning. Jason, Lynn, and Dr. Munyard have enthusiasm and passion for this testing and what the results can do for improving breeding programs and fiber quality for alpacas worldwide. Lynn's latest crop of crias continues to demonstrate the power of this testing and how it can improve fleece characteristics and color outcomes.

Dr. Munyard continues her research and is actively working on expanding this initial information. We expect to see much more coming on color genotype testing and should have even more data to work with in future years.

Example Color Outcom

Estates Sakima's
Braxton
EEaa

Cannot Produce
Offspring With
Light Coloring

Snowmass
Occult
Eeaa

Can Produce
Offspring With
Any Coloring

r Different Herdsires

Snowmass
Moonlight Lover
eeAa

Can Produce
Offspring With
Any Coloring

Snowmass
Coalition
eeAA

Cannot Product
Offspring With
Black Coloring

CHAPTER 10

OPTIONS AND TECHNIQUES FOR BREEDING ALPACAS

Breeding alpacas is an amazing experience and one I am always thankful to participate in. Each pairing brings possibilities of what might be created, and each birth brings new life. I've also realized each breeding produces additional opportunities for me to strengthen my bond with the dam and create a new relationship with her offspring.

I always feel a sense of magic and peace when we deliver a healthy baby. But it's also important to understand that not everyone is cut out for the life of livestock breeding. It takes time, money, patience, and a willingness to make difficult decisions and do hard things.

Breeding Considerations

Before you decide if raising and breeding alpacas are right for you and your farm, I'd like you to consider some of the challenges involved and the qualities that will help make you successful.

Time

First and foremost, you need to have enough time to allocate to the process of learning, analyzing data, breeding, rebreeding, checking on pregnant females, birthing, and sometimes bottle feeding. Cria watch (explained in the next chapter) alone is an activity that you do every hour during daylight for anywhere from two to four weeks. You must have enough time in your schedule (or have a caregiver) to accommodate the care of the alpacas.

Some Medical Knowledge

Having a medical background or experience in farming will help make the entire process easier. I came into this as a city girl with zero knowledge. My husband grew up helping his uncle raise hogs in the summer, so he had a basic knowledge of farming. Neither of us came prepared for breeding alpacas, so we relied heavily on our mentors Nancy and Kim. We've learned a lot, and we take good care of our alpacas, but this required us to quickly jump in and become proficient in the nuances of alpaca medical care.

I bought every book I could find on raising and breeding alpacas. I've read them, highlighted text, put bookmarks in them, and had them in the paddock with me during deliveries. Since I didn't know a lot to start, it was my responsibility to do what I could to educate myself on what I was missing. Thankfully the alpaca industry has some amazing vets and educators who also provide seminars on breeding and raising alpacas. If you plan on breeding alpacas, know you must educate yourself on the process. Have a plan in place (like reading books, attending seminars, or consulting mentors) to learn about the medical care needed for breeding, birthing, and cria care.

Physical Space and Shelters

In earlier chapters, I've touched on the need to keep male and female alpacas in separate spaces. In case you missed this, I would like to reiterate that again and suggest you review Chapter 5 on farm setup. And, as another reminder, don't forget you'll need an additional shelter and paddock area for young males.

Ability to Perform Medical Tasks

Breeding will bring the need for additional medical care. From checking the status of a hymen on young females to adjusting crias inside the birth canal, there will be times when medical care is needed. Sometimes a vet can do this for you, and sometimes urgency will dictate that you jump in and quickly manage the situation yourself. Medical tasks are not for everyone, so go into breeding knowing this is a requirement.

Money

While breeding alpacas can be lucrative, you need to have enough money available to buy a starter herd and have a fund for unexpected emergencies. Unplanned medical issues will happen when you're breeding and delivering these unique creatures. For the most part, alpacas can birth on their own and without issue. However, there are times when you need a plasma treatment or something else that requires a professional alpaca vet or teaching hospital.

Compassion

Alpacas are highly intelligent animals, so they need someone who will be empathetic, calm, and patient. My compassion for their late pregnancy angst, my empathy for their pain during contractions, and my patience with daily visits to build trust have made birthing easy for me and my herd. The alpacas can sense my compassion, which helps me effortlessly slide out a stuck cria, get milk started for a newborn cria, or perform medical care on their babies. When your alpacas trust you, you can focus less on the dam's reaction and instead focus entirely on the medical task at hand.

Ability to Say Goodbye

I mentioned already that I cry when alpacas leave, and I do. Every single time. And this is even after I've found highly qualified buyers to make sure our alpacas go to good homes.

I cry because I love them. If I could, I would keep all of them. I know I can't, so I do my best to muddle through the emotional turmoil of the sales process.

When planning on breeding, you should go into it with the expectation of selling alpacas. You cannot breed and collect hundreds of animals, so you must be able to say goodbye and sell them when needed. For some of us with sensitive personality types, this is more difficult than you might expect.

Familiarize Yourself with the Basics of Breeding

There are a lot of people who assume alpacas are bred and cared for like the family dog or other livestock animals. This is a false assumption. Alpacas are unique, and you cannot go into this process thinking you'll breed alpacas like you'd breed a Labradoodle.

If you want to successfully breed and raise alpacas, you need to know all about their nuances. These include both biological and emotional elements. I'll run through my favorite alpaca tidbits on the process of breeding.

Female Age of Maturity

While female alpacas are generally sexually mature between ten and fourteen months of age, alpaca experts recommend holding off on breeding until the female reaches fifteen to eighteen months and a minimum of ninety pounds. A female alpaca's ovarian activity doesn't reach its fullest potential until about seventeen months of age.

On our farm we hold off breeding until a female is at least two years old and she demonstrates she is sexually mature with a desire to breed.

Male Age of Maturity

Male alpacas can begin breeding and successfully impregnating females between thirteen and twenty-four months, although there have been cases of much younger fertility in males, as well as late bloomers who are not ready until twenty-four to thirty-five months. Each male is different, and you should have an experienced vet assess physical and mental readiness.

Induced Ovulation

Unlike human females, alpacas do not have a "monthly cycle"; they begin the ovulation process once the mating process has already started. Ovulation typically occurs twenty-four to forty-eight hours after mating. Keep in mind that for ovulation to be effective, the female needs to be receptive to the male and in a low-stress environment.

Alpaca Penis

The alpaca penis has often been referred to as a corkscrew. It is long and thin with a slight curve. The corkscrew portion is a tip that is used to dilate the cervix and enter the uterus during breeding. While this physical property works great for breeding the ladies, it can cause significant damage (and even death) to a female alpaca if breeding takes place too often. This is the reason we keep males and females completely segregated and give the ladies time to heal between breedings.

Castrating a male (also called gelding) does not change the shape of the penis or the ability to enter a female. A gelded male can still form an erection, penetrate a female, and do significant harm to the female internally.

Cushing

If the female alpaca is interested in breeding, she'll lay down and allow the male to mount her. We refer to this position as cushing. If the alpaca has no interest, she'll spit, run away, and avoid all interaction.

Orgling

When you bring an experienced male into the female pen, he'll puff his chest out, prance around, and start orgling. The orgling sound is hard to describe, but once you hear it, you won't forget it.

Leviticus is our resident stud, and all the ladies love him. One of the main reasons the ladies love him is because he is the master of orgling. He starts his music, the ladies drop, and a baby is made.

The Alpaca Spit Test

If the female is already pregnant, she'll let the male know by aggressively spitting in his face. Kalypso, an experienced female breeder, will literally run to the male, throw a huge gob of spit at him, and then turn around and leave. We refer to this behavior as a "positive spit test." And as surprising as it may sound, this spit test is about 80 to 90 percent accurate.

Alpaca Gestation Period

A female alpaca is pregnant for about 11.5 months. Data has shown the average length of time ranges between 335 and 355 days, with reported pregnancies going out as far as 380 to 400 days. Our personal record is Kalista at a whopping 370 days.

Different Techniques for Breeding

The alpaca industry uses a variety of techniques for breeding. The best technique will be heavily influenced by the size of your alpaca herd. Following is an overview of possible breeding techniques to use.

Halter Breeding

Smaller farms like ours will halter breed (also called hand breeding), which means we will bring a male on a halter to the female pen and point him toward the lady we'd like him to breed. We match males and females up based on their color genotype, histogram data, or EPD information (see Chapter 9).

The male will breed for about twenty minutes (the maximum time suggested), and once he is done, we bring him back to his pen. We wait about seven days to bring him back to see if the female is still interested. If she is open to breeding, we do so. Otherwise, we assume the female is starting a pregnancy and we return the male to his pen.

The benefits of halter breeding are that you control the duration of the breeding session, you'll know the exact date of a successful mating, and you'll be able to estimate an accurate due date for the cria.

Pasture or Field Breeding

Pasture breeding is a technique used at large alpaca ranches with hundreds or thousands of animals. When pasture breeding, you leave a male in with a large group of females (about thirty) for about a month. He'll

Coalition is breeding Adel, and Diva is watching and waiting for her turn. Diva's behavior is a clear sign she isn't pregnant, but is easier to breed.

go through the ladies and do his work until all of them are pregnant. An experienced male is smart enough to know who is open (available), who is pregnant, and who needs to be left alone.

Pasture breeding offers the benefits of a natural setting, the male can breed at his own pace, and there is limited effort on the part of the humans. There are also a few negatives. You must have a large number of females for the male to work with, you won't know the exact data of a successful breeding, and your estimated due date will be unknown, which means you'll be on cria watch for an extended period of time.

Drive-By Breeding

A drive-by breeding occurs when you're using someone else's male to breed. You load up the male or the female and drive them to the destination farm where the breeding will take place.

It's pretty easy to do a drive-by breeding with males, and this is especially true if they've mated in this format previously. The boys know they're headed to a date, and they are eager to perform once they arrive at the destination farm.

Eddie came to perform a drive-by breeding with Vin. Eddie is always a gentleman and even brings his own snacks.

The ladies are a bit different. Some ladies are perfectly agreeable to drive-by mating, while others are just too stressed by the location change and not receptive.

Remember That Every Alpaca Is Different

Over the years of breeding alpacas, we've learned that each alpaca is different. You cannot expect to breed the entire herd in the same manner. You might be able to do so with other livestock, but not with these lovely creatures. They are too smart, and they have way too much personality to expect generic behavior from them. If you want to maximize your success, you need to know your alpacas intimately and breed them according to their unique physical and emotional traits. Here are some examples of different alpaca breeding scenarios we've encountered on our farm:

- **Faith:** We tried multiple drive-by breeding attempts with her, but I finally called it quits until we could breed Faith at our farm, in her paddock, and on her terms. As soon as we switched to this process, Faith was happy to breed and she became pregnant quickly.

- **Sherry Ann:** This lady had a stellar pedigree and we were eager to breed her. Unfortunately, when she was physically ready to breed, she wasn't

mentally mature enough for the process. Sherry Ann displayed zero interest in the male, which was a clear sign she wasn't ready for breeding.

- **Diva:** When Diva was old enough to breed, her weight wasn't at the level needed. We had to postpone breeding until months later when she was fully ready.

- **Lilly Grace:** This young lady is old enough, large enough, and mature enough to breed. The problem is she is a "roller." Instead of staying in a cushed position, Lilly Grace will roll around under the male. An inexperienced male will just get up and walk away, so we need to use an experienced male to keep her in place so the semen can transfer successfully.

- **Romeo:** Romeo was a late bloomer, and he just couldn't quite figure out the breeding process at first. He knew he liked the girls, but for some reason he didn't understand the concept of putting his butt down. Eventually, he figured it out, and he has produced multiple babies as a result. He was totally fine with all the trial and error, but my husband and our lady alpacas found the process extremely frustrating.

- **Leviticus:** Leviticus brings all the macho energy you could ask for. But this comes with a cost. When we halter breed, Jason must manage him. At five feet nine inches tall, I'm not large enough to manage Leviticus. He isn't aggressive toward me, but he is very active and fast. The boy is just overly excited when it's time for a date, and his energy level is more than I can physically control.

- **Ariana:** The majority of the ladies tend to be "one and done" females. This means we breed them once and they are pregnant. Alpacas tend to be eager breeders, and when this is the case, pregnancy happens quickly. Ariana is a one-time girl. Get her the right male, she'll eagerly breed, and she will be pregnant.

THE LADIES ARE IN CHARGE

To perform successful and healthy breeding sessions, let the female alpaca determine when and if she breeds. Never force a breeding match or restrain a female during breeding. It's inhumane and it won't produce a successful outcome.

Use Data to Match Your Breeding Pairs

When people contact us for alpacas to buy and breed, the first thing we ask is what their herd is lacking or what their breeding objectives are. A lot of people think it's odd that we ask these questions, but we do it to help make sure the farms are breeding for what their herd needs.

When we are ready to start the breeding season, our family sits down together to discuss our objectives for our own farm, the herd as a whole, and each alpaca. Do we need to improve fineness, density, staple length, crimp, or uniformity? Or maybe would we like to work on shifting the colors produced?

Once we know what we need to improve, we match criteria to specific alpacas that carry those traits. This trait data originates from histograms, EPDs, and visual inspections.

EXAMPLE BREEDING MATRIX

Herdsire	Male Positive Traits	Female Traits to Improve
Fun in the Country's Leviticus	Producer of dark offspring Long staple length	Short staple length
Snowmass Occult	Beautiful "show" head Strong crimp structure Producer of dark offspring	Limited facial fiber Low curvature in crimp
Snowmass Coalition	Low micron Strong uniformity Dense fiber High-frequency crimp Low medullated fiber	Bolder crimp Higher micron Higher standard deviation Low mean curvature
Snowmass Moonlight Lover	Beautiful "show" head Long staple length Strong crimp structure Producer of dark offspring	Limited facial fiber Low curvature in crimp Short staple length

By matching up individual alpacas, you not only improve specific breeding outcomes but also your herd as a whole. When breeding alpacas, we always want to "breed up" so we can continuously improve our herd and alpacas throughout the industry. If we use the data available, we can and will improve the quality, revenue generation, and health of the animals. While many hobby farms will need to take a simpler approach, if you'd like to create a long-term breeding program, the industry data can help you make better decisions and have a long-term impact on the quality of your herd.

INBREEDING VS. LINE BREEDING

Line breeding is the process of breeding *distantly related* alpacas to increase the potency of specific traits. Inbreeding is the process of breeding *closely related* alpacas, such as father to daughter, mother to son, or brother to sister. Line breeding produces strong characteristics, while inbreeding produces significant health issues, abnormalities, and at times death.

The Alpaca Owners Association offers data on the "coefficient of Inbreeding," or COI, for every registered alpaca. COIs measure the genetic similarity over generations of an alpaca's pedigree, which helps prevent situations of inbreeding.

CHAPTER 11

Delivering Healthy and Happy Cria

I love alpaca babies and the entire birthing process! It is a beautiful, mostly peaceful process that fills me with awe and wonder. Jason is the exact opposite. He is so worried about taking good care of the dam and her cria that birthing season fills him with endless anxiety. He is restless and struggles to manage the births because his stress rolls over to the dam and her cria.

With each year we birthed babies, Jason's anxiety rose, and we could see the negative impact it had on the birthing and aftercare. We made a conscious effort to shift birthing responsibilities from him to me.

I headed to Ohio State University to take a wet lab and neonatal clinic. I walked away from that clinic feeling knowledgeable about birthing, but even more important was the fact that I felt *empowered*. It was a game changer for me and our farm. I was no longer afraid of what to do. I knew what to do and didn't hesitate when faced with doing it. We went from having tons of birthing problems to a year of effortless births and excellent baby health.

I want to help you by offering some of this knowledge to get you started.

This chapter will cover what to include in your birthing kit, what signs of labor to watch for, what to do during the birthing process, and what to do after the baby arrives. While this content isn't a replacement for a neonatal class, it is enough to provide insight into what to expect and how you can contribute to the birth of a happy and healthy cria. I hope this overview helps you feel empowered so you can experience the same wonder that fuels me each spring.

Love & Grace with baby Zachary.

DO NO HARM

The primary rule of birthing alpacas is "do no harm"—something every alpaca owner must remember. As a breeder, your role is to step back and let the dam naturally give birth. If progress doesn't occur, then and only then do you intervene.

Your Presence Can Make All the Difference

As the weather warms and spring arrives, I spend more and more time in the barn with our pregnant girls. I want to closely monitor their pregnancies, but I also want them to know I'm present and they can trust me. This is especially important for a first-time mother or a female we've bought from another breeder.

Spending time with the ladies can mean anything from sitting out with them in the paddock to reading a book in the barn. I just want to be present, interact with them, and remind them again and again that I am safe.

Last year, we bought Snowmass Royal Favor and brought her home mid-pregnancy. She came from a very large farm and was not used to having a human hover around her. I persisted and went out of my way to interact with her. It felt like she was tolerating me at best. But I was so wrong. When active labor hit and I bent down to talk to her, she started licking me all over my face. She had true love in her eyes while doing so.

You can take that action to mean her endorphins were kicking in and I was receiving some early love meant for her cria. Or you can assume my efforts worked and she was thankful I was there with her. Based on her expression and demeanor, I believe she was thankful.

Midway through the delivery she seemed to give up. The cria's head and front legs were out, but Royal Favor appeared to be out of energy. It was clear I needed to help, so I washed my hands quickly, put plastic gloves on, and rubbed lube all over my hands and arms. While I bent down near the baby and calmly talked to Royal Favor, I investigated the situation to determine exactly what I needed to do. Within a few more minutes, she and I wiggled eighteen-pound baby Ben out into the world.

Throughout the process Royal Favor was calm, and what could have been stressful and chaotic for both of us was peaceful and beautiful. Together we had an amazing birth. Ben is now a big, healthy boy, very active and handsome. And what really makes me happy is that the bond I now have with Royal Favor is strong and unwavering.

What to Include in Your Birthing Kit

Having a well-stocked birthing kit on hand is key. Not only will it help you protect your dam and cria, but being prepared will also reduce stress should a crisis occur. Your kit should include:

- A large quilt or blanket to lay the cria on immediately after birth
- Multiple bath towels for cleaning and drying off the cria
- A blow dryer to expedite the drying process if needed
- A container of Clorox wipes (to clean your hands or tools quickly in an emergency)

- Sterile gloves and lubricants (if you need to intervene, you will use a lot of lube) for repositioning a baby

- Scissors and dental floss for addressing umbilical cord or afterbirth issues

- A container with 7% iodine diluted with water (50/50 mix) for dipping the umbilical cord

- A thermometer to check the cria's body temperature

- A bulb syringe to clear the cria's airways if necessary

- A livestock scale to weigh the baby

- Emergency colostrum and a bottle in case the cria needs supplemental feeding

- Frozen llama plasma in the event failure of passive transfer occurs

- Banamine and syringes for giving pain relief (provided only after the placenta passes)

- Cria coat (used during colder weather deliveries)

We keep our cria kit in a large plastic tub inside our med room. When we go on cria watch (two weeks before the estimated due date), we pull out this kit and set it in the corner of the ladies' barn so I can quickly grab it and access materials as needed.

Baby Otto feeling healthy after his plasma treatment. Our local vet will come to our farm and perform the plasma treatment right in our store so we have a controlled and clean environment.

ORDERING LLAMA PLASMA

Keeping a bag of plasma on hand is important. When this is needed, you have only twenty-four hours to administer it, and most vets will not have it on hand. It will be your responsibility to source and store this item in advance. The alpaca industry generally uses llama plasma, which originates from Triple J Farms in Bellingham, Washington. Useful Llama Items, Inc. keeps this in stock and is one of the most economical places to order it.

The Baby Is Coming!

You've waited over eleven months for the cria to grow and you're anxious for the upcoming delivery. But how do you know when your alpaca is ready to begin labor? Are there clear signs an alpaca is ready to deliver?

Over the years I've been making notes as we've birthed our crias and documenting any signs I noticed, so that I could determine if there are some that tend to trend across the herd. While my list is not scientific, it is based on my own observations and hopefully will be helpful for new alpaca breeders to ascertain if delivery is imminent.

Cria Watch: Two Weeks Before Delivery

Cria watch is a term we use to define the time frame between two weeks before the due date and the day the baby arrives. This two-week period might be exact with the anticipated due date, or it might end up being two weeks later than you expected. During this time, we check on the dam throughout the day to assess her personality and physical changes. Across the alpaca industry, farms and vets have noticed a distinct behavioral change about two weeks before birth. The female will act as if she is in active labor and the baby is arriving soon. You'll question this because you know it isn't time yet. The dam will moan, roll around, and mimic many of the same movements of an active delivery.

Delivery is *not* coming at this point, but the baby *is* moving into position. Take note of this change, mark your calendar, and expect a baby to arrive two weeks later.

For smaller females, this behavior tends to be more pronounced. One year, our pint-size Dolly was so uncomfortable we loaded her up in my SUV and rushed her to our alpaca vet to make sure we didn't have a uterine torsion (rotated uterus) on our hands. We did not. Dolly was just struggling with the baby changing positions due to her small size.

With drama queens, these movements are also very noticeable and memorable. When Kalista was pregnant with Walter, she threw herself on the barn floor and moaned as if she was dying. At first we thought she was having a difficult delivery, then we realized she was being a bit excessive as the baby moved into position.

The Days Leading Up to Delivery

In the days leading up to the delivery, signs of labor will be different from female to female. Some alpacas will have lots of birthing signs and some will have no signs at all. For the ladies that do show signs, here are the common ones we tend to see:

- **Enlarged teats:** For some females, their teats will begin filling up with milk, and if you look, you'll be able to see this drastic change. This is harder to see if the female is black or in full fiber.

- **Puffy butt:** "Puffy butt" is a nonscientific term we use to describe an enlarged back area. In some females, this will pulse and go in and out. We've noticed it tends to be more prevalent in the early evening.

- **Perineal relaxation:** This can start weeks or days before the delivery, at which point it is only slightly noticeable, and it can rapidly change in the hours leading up to the delivery. Perineal relaxation seems to change based on the position of the alpaca's body. It is more evident if they are lying down, so don't let this throw you off. Wait until the dam stands.

- **Tail up:** About half of our alpaca ladies will hold their tails up in the few days prior to delivery. When

Dolly's "puffy butt" is a clear sign we'll be having a baby soon.

the tail isn't held up, they seem to be whipping it around. I have no idea why they do this, but a decent percentage of them present this behavior.

- **Baby shifting position toward the back:** If you keep a close eye on your female alpaca's back end, you'll see things shifting as delivery comes close. At times, it will look like there are baby toes trying to poke their way through the skin and fur.

- **Moaning:** Some ladies suffer in silence and others moan constantly. For the more vocal gals, you'll hear a lot more moaning in the last day or two. It can come and go, or it can be a constant stream of complaining.

- **Increased eating:** Increased eating a few days prior to delivery seems to happen with almost all of our pregnant alpacas. It seems no matter how much food you give, it is never enough. I launch into Italian grandma mode and make special accommodations for these ladies. They know exactly what I'm doing and eagerly join me for special snack time.

As you have births year after year, you'll see alpacas tend to repeat their specific signs for each birth. I know Dolly will eat and drink nonstop, Kalista will act like the world is ending, and Royal Favor will become quiet until active labor arrives.

The Day of Delivery

Most alpacas give birth in the morning hours, often on sunny days, which is a natural adaptation to ensure warmth for the newborn cria. The good news is you've typically had your cup of coffee but you're not yet ready for dinner. As the delivery approaches, there are some clear-cut signs that the baby is coming soon:

- **Look of fear:** Many first-time dams will have a look of fear as the birth is imminent. If you don't know your alpacas very well, you might not notice this change. If you're close with your ladies like I am, you can see this fear take over and become visible on their little faces. I tend to talk to them a lot at this point. The ones who really love me welcome it. The ones who tolerate me look at me with disgust.

- **Lying on their side:** As contractions start, the alpaca will begin to lie on her side a lot. She will switch between this position, a traditional cush position, and standing.

- **Even more perineal relaxation:** I mentioned this one above and I'm discussing it again because there is a distinct difference at this point. While it can start many days in advance, it takes on a whole new look when active labor arrives. I sat with my favorite girl Nibbler the entire day she delivered her first cria. I could see this changing as the hours went on, and sure enough, when it seemed nice and ready, she dropped out her baby in a matter of minutes.

Nibbler's perineal relaxation.

- **Lost interest in food:** My husband's favorite sign of delivery is based all around food. If he goes out to do morning chores and one of our pregnant alpacas doesn't run to her grain bowl, he knows we have a baby coming. And for the most part, he is generally right. When they stop eating, the babies start coming!

- **Excessive water intake:** When we've had hot, dry summers, we notice some of our pregnant alpacas become excessively thirsty. This sign is a strong signal the baby is coming. If you see your pregnant female visiting the water bowl a lot, take note. There may be a strong chance she'll be starting her labor and delivery that afternoon.

- **Trying to poop a lot:** I have about a dozen books on alpacas and all of them reference increased pooping (or trying to poop) as a signal of birth. Yes, it happens, but it doesn't happen for every alpaca. We watch for this sign but also know it isn't guaranteed.

- **Scratching their stomach and back:** As contractions start gearing up, so does the scratching of the stomach and back. You may even see the dams rub against a fence or hay feeder. I always wonder if the pregnant alpacas feel they have a fly on

them when really, it's just a contraction. Regardless, I have found this to be an increasing behavior as delivery approaches.

• **Separation from the herd:** When we are reaching the end of delivery preparation, the ladies are tired and anxious. Due to this, many of them (but not all) will opt to go off and leave the area of herd activity. Our pregnant ladies still stay close enough to see what is going on, but far enough away that no one can bother them. Regardless of the dam's wishes, the herd will follow and get all up in her business when they see a baby coming out! I often have to push them back so the dam can deliver without everyone crowding around her.

Immediately Before Delivery

By the time active labor is happening, I've been anxiously watching for days. I know every moan, every sign, and I've had long talks with the girls about what is coming. I always find myself telling them how excited I am to meet their new baby. Yep, I'm that type of nutty alpaca owner! I can't help it. I had a very difficult birth with my own daughter, so I have lots of empathy for the alpacas and I feel compelled to try to calm my alpaca girls before delivery.

I have three remaining delivery signs for you to know when labor is active and delivery is imminent:

- **Panting:** When you see your alpaca panting like a dog, it's time to grab the cria birthing kit. The baby is in the birth canal, ready to make its grand appearance.

- **Grunting:** As the baby starts to approach, moaning turns to grunting. Just like we human ladies who grunt their way through delivery, alpaca mommas can and will do the same.

- **Digging:** My final sign is the one that tends to happen minutes to seconds before I see the cria's nose and legs stick out. The pregnant dam will start digging in the dirt, grass, or straw. I often wonder if she is looking for a soft location for the baby to drop into.

If everything goes as it should, you'll see a nose and then the front legs. The baby will look like Superman flying out and arriving in grand style. Some dams deliver lying down, but most deliver standing up.

Once the cria is delivered, the mother tends to walk off and leave the new baby with the other female alpacas. This is okay. Many alpacas will pass the placenta well away from their baby. We suspect they do this so the placenta doesn't encourage predators near their baby. Once the placenta is passed, the mother will return to start bonding with her baby.

Adel giving birth with Jalapena investigating the baby.

The really good mothers will nudge the cria, love on it a bit, and then stand like a statue as the baby finds its feet and attempts to nurse. It can take some babies awhile to figure everything out, but once they do, a good mom will stand ever so patient while the baby takes in milk.

Timing Your Deliveries

Alpaca births typically proceed smoothly without human intervention, but it is important to monitor the process closely and watch for issues. The birthing process occurs in three stages:

- **Stage one:** The dam will show signs of discomfort, such as pacing, humming, and lying down and getting up frequently. This stage can last anywhere from one to six hours as contractions prepare the cervix for delivery.

- **Stage two:** At this point, active labor begins and the cria is delivered. Delivery usually happens within thirty to sixty minutes. If the process takes longer or the cria is in an abnormal position, assistance may be needed.

- **Stage three:** After the cria is born, the placenta should pass within a few hours. If it does not pass within four to six hours, veterinary intervention may be necessary.

It's important to note that you cannot give any painkillers until the placenta passes, because the medicine will prohibit the expulsion of the placenta. No matter

THE FIFTEEN-MINUTE RULE

Once you start to see a baby, closely track the progress of the delivery. If fifteen to twenty minutes passes without advancing the birth, you might have a dystocia birth. This means the baby is not in the proper position and you will have to assist. This is when a prior birthing class is very helpful!

I take photos of the toes, head, and shoulders as they become visible to help me time the births. I text them to my husband to give myself quick time stamps for each image. This serves as a sanity check so that I know exactly how much time has passed. It is easy to be caught up in the process and think thirty minutes have passed when really it's only been ten minutes.

how much the dam might need them, you must wait until she passes the placenta. The good news is many times no painkillers are needed. We generally see the need for these with maidens (first-time mommas) or births where we've had to assist.

What to Do After the Birth

When that baby finally arrives, you're going to want to jump in and assist. I beg you to stop and let nature take its course. Truly, the best thing you can do is step back and wait for the dam and cria to emotionally connect first. They need to form that initial bond, and you should not interrupt that process unless the cria or momma are in dire need of help.

When my husband was doing the births, he would jump in immediately. Back then, I was still 100 percent a city girl, so I just let him do it. I didn't have confidence with the birthing process and I deferred to his judgment. After a few years of watching this process and our alpacas having lots of trouble with post-birth milking, I finally stepped in and insisted on some changes. It took me awhile to convince my husband that our desire to help was hurting the natural birthing and bonding process—it was greatly interfering with the dam's ability to bond with her baby. I finally convinced him to "try it my way," and when we did, everything changed for the better.

What you want to do is let the dam and cria emotionally connect. You will know when it happens

Dolly bonding with her newborn.

because you'll see and hear this bond start to form. The baby will whimper, and the dam will cluck, hum, or begin licking the baby. Stay quiet and let this interaction occur. Once you see the baby knows who mom is and mom knows who baby is, you can begin to dry off the baby. It is very rare for a dam to clean off the newborn herself.

When you approach the cria, stay low, make eye contact with the mother, and talk to the mother calmly so she knows you are there to help, not to steal away her cria. Go slow and be reassuring. This will keep her stress down and help create a positive, calm environment for the baby.

Once the cria is delivered and that emotional connection is in place, there are several important steps to take to ensure the health and well-being of your cria:

- **Check the breathing.** Make sure the cria is breathing normally. If necessary, use a bulb syringe to clear the airways.

- **Move mom and baby into a protected area.** We have a cria area in our barn that is big enough for mom and baby to safely move around, but separate from the herd so the nosy aunts don't interrupt. We'll keep mom and baby here for two days. This allows sufficient time for them to bond and learn about milking. New babies cannot control their internal temperature for the first few days, so this segregated area also helps keep the baby from overheating or becoming chilled.

- **Dry the baby off.** Use clean bath towels to dry the cria and ensure it is warm, especially in colder weather. I go through at least two or three towels during this process.

- **Dip the umbilical cord.** Use an iodine mixture to disinfect the umbilical stump and prevent infection. You'll want to do this three times over the first twenty-four hours. We put the solution in a short, plastic shot container and bring it up to the baby's cord.

- **Monitor the baby's movement and standing.** The cria should attempt to stand and nurse within the first thirty minutes. Some babies will take one to two hours to figure out their legs. If the baby is strong, stay out of the way. If the baby is struggling, provide assistance. As noted before, talk to the dam so she knows you are helping her and the baby.

- **Record the weight.** Weigh the cria and record the weight. Document the time of birth and any observations. I do this in a Notes app on my phone, and my husband then records the information on paper for our files.

- **Get that baby milking.** Ensure the baby gets up, knows where the milk is, and gets a taste of it. The baby needs to drink a good amount of milk in that first twenty-four hours so it obtains colostrum and the vital antibodies the dam passes on. If this does not happen, you'll have a situation called "failure of passive transfer," and you'll need a vet to administer plasma via IV to save the baby.

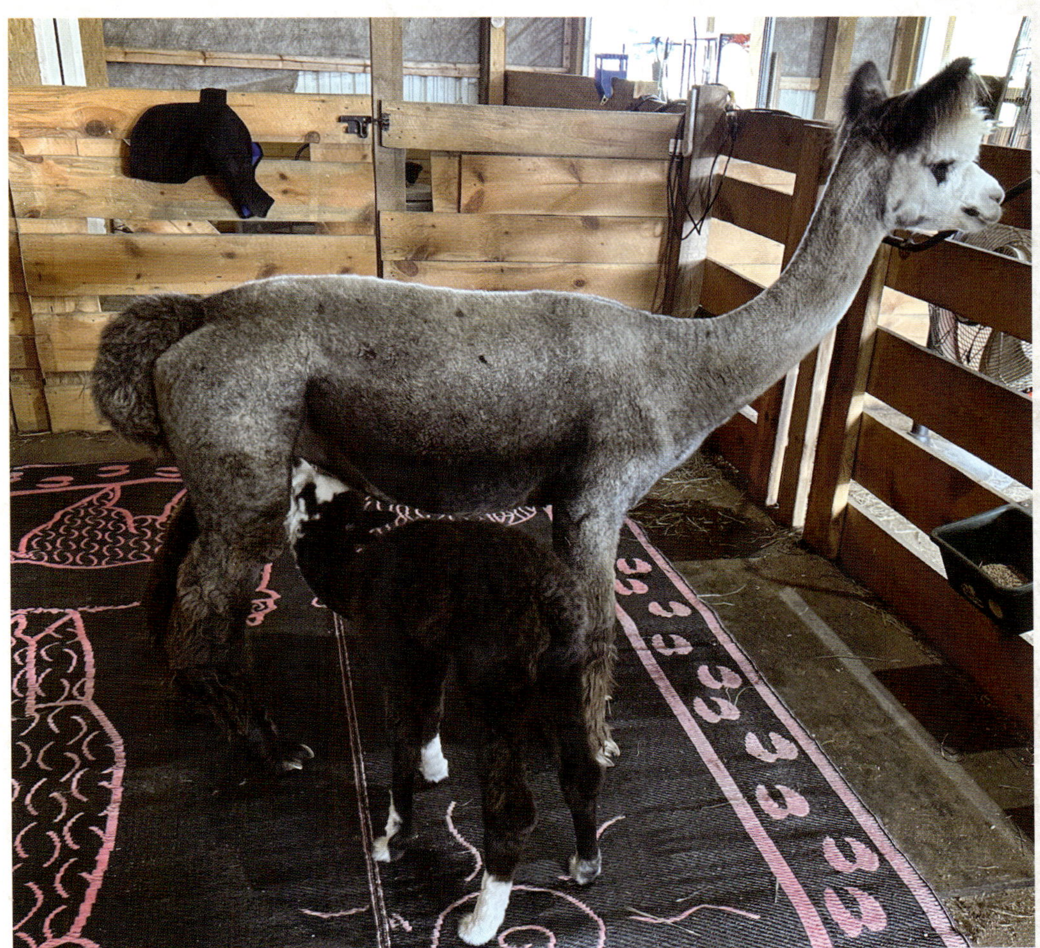

Divo is just starting to understand milking. His mom Diva is being a rock star by standing like a statue to allow him to nurse.

Sometimes you do need to step in and help the baby find the milk.

- **Observe the dam.** Check for proper placenta expulsion (this should happen within a few hours) and monitor her for any signs of distress or complications. If the dam is in pain, contact your vet for guidance on treatment. As a reminder, painkillers cannot be administered until the placenta has passed.

Once I've performed this initial check, I will do a lot of nothing—and this is by design. I praise the dam for her beautiful baby, then I simply stand back, watch, and enjoy the miracle.

If the baby doesn't seem to be "getting" the milking process, I will continue to assist. I've had dams reward me with a peck on the cheek, because they knew I was helping their babies. As long as the baby knows where both mom and milk are located, I will exit and watch remotely on our camera system.

If at any time you feel anything is wrong, call your mentor, vet, or a teaching hospital. You have twenty-fours to get this baby stabilized, so don't delay!

CHAPTER 12

WHAT TO KNOW ABOUT ALPACA SHEARING

If you're wondering how alpaca shearing is done and what you should expect during your first shearing day, you are not alone! I was pretty darn confused our first year of owning alpacas. In this chapter, I'll walk through how we handle shearing so you can get a good feel for the process, planning, and execution. As I said in Chapter 3, alpaca shearing is not for the faint of heart. It's part art and part dexterity, which is why most alpaca owners choose to hire a dedicated, trained shearing team.

In the United States, alpacas are sheared each spring, which produces the fiber (also called fleece, fur, or wool) that represents that year's harvest. In their native Peru, alpacas are sheared every eighteen months. Peru has a more consistent temperature, so locals can extend the duration of growth while still keeping their alpacas comfortable and safe.

Shearing Is Harvest Day for Alpaca Owners

Shearing day is a major event at any alpaca farm! It not only prepares the alpacas for the summer heat, but also is an opportunity to review the quality of an individual alpaca's fiber. And once removed, the fiber will be converted into products, which makes this process the first step of revenue generation for many alpaca owners.

An adult alpaca weighing 150 pounds will produce about 5 to 10 pounds of fiber. This amount will vary based on the alpaca's physical size, the length of their fiber, and the density of their fiber blanket. Fiber density and length play an important role in the amount of fiber grown. For instance, Dolly is a pint-size female on our farm and our smallest adult female. The interesting part is she is a top producer of fiber and always has been due to her great staple length and density.

Once removed, the raw fiber will be graded for quality, cleaned, and then converted into yarn, clothing, or household items like throws and dryer balls. The final destination of the fiber is determined by the micron count (fineness) of the fleece. The lower the micron count, the higher the quality of fiber. A standard alpaca produces approximately four pounds of high-quality fiber (we call these firsts) and an equal amount of shorter/coarser fiber (we call these seconds and thirds).

Shearing Day Can Be Draining and That's Okay

Shearing an alpaca is not for the faint at heart. In our first year of shearing, I was totally unprepared, and I had to walk out of the barn to get my emotions in check. I simply didn't realize that shearing a herd of alpacas is like taking a group of human toddlers to their first dentist appointment.

This is me and my favorite girl Nibbler before shearing. She struggles with anxiety on a regular basis, and I wasn't sure how she would do in her first year of shearing. We were both really nervous, but I did my best to hide it. (This photo is from 2020 during Covid, so I'm in a mask and both of us are in desperate need of a haircut. Thankfully, you would normally not wear a mask for shearing.)

By year two I was perfectly fine because I understood what was going to happen and how specific activities (like restraining the alpaca) were done for the health and well-being of the animal.

Prepare Yourself for the Big Reveal

Patrick is Nibbler's cria and the photos shown here are from his first shearing. Notice how Patrick has different colors between his baby fur in the photo before shearing and the color of the fiber in the photo after shearing.

MOLLY
Before & After Shearing

Where did those ears come from?

Before & After Shearing

Different color, but still adorable!

This change in color is because the fiber tips were modified by the amniotic fluid in his mother's womb during gestation. When an alpaca is born, we really don't know what color their fiber will be until it starts to grow outside of the womb. Removing these cria tips is always exciting because you may find a completely different-colored alpaca hidden underneath!

Is Alpaca Shearing Cruel?

You may have heard people say alpaca shearing is cruel. It is not. More important, it is done for the safety and welfare of the animal. I cannot stress that enough.

Organizations like PETA would love you to think alpaca shearing is inhumane and wearing alpaca fiber is ethically wrong. That is 100 percent false, and I get very frustrated with animal rights organizations that misrepresent our industry and miseducate consumers.

Alpacas do not shed their fur like a dog or cat. In the United States, they must be sheared annually for their own health and welfare. *Not* shearing them is what is cruel because they can't manage the summer heat and would die of hyperthermia.

My husband and I exercise caution and close oversight when managing shearing day. We hire professional shearers, and we are with our herd the entire time that shearing is taking place. We know exactly what is done and why it is done, and we make sure nothing cruel ever occurs to the alpacas. We also use shearers that we know will take even greater care with our pregnant girls and those who need special handling due to age.

Why You Should Consider a Professional Shearing Team

Most alpaca owners will hire professional shearing teams so they can guarantee their animals are moved through the process quickly and efficiently to reduce stress on the animal. And most alpaca owners also stay close to assist the team and increase the efficiency of the process. We do so because we truly care for the welfare of our livestock and want to give them the love they give to us.

Here are the reasons why we opt for professional alpaca shearers:

- The process requires specific equipment to help ensure the safety of both the animal and the human shearing them.

- Having a specially trained team makes the process more efficient, quicker, and produces a lot less stress on the alpacas.

- It's a three- to four-person job that takes strength and dexterity.

- While the alpacas are sheared, the team will also trim teeth and clip toenails.

- A professional shearer will preserve as much of the wool as possible, reduce cuts within the fiber, and create a more uniform fiber blanket for converting into goods.

Professional shearing is not cheap, but the money spent is well worth it.

How to Get Ready for Shearing Day

On our first shearing day, we were clueless. We didn't know what to do to prepare or what to expect. What we did learn was invaluable. It boils down to a few steps that make the entire process go faster and a lot smoother for everyone involved:

- Keep the alpacas as dry as possible. If you need to lock them in the barn the night before, do so.

- Remove any straw you have on the barn floor. This makes shearing easier for the professionals and cleaning the fiber much easier for the owner or mill.

- Vacuum or blow out the alpacas' fiber the day before shearing. The more vegetation you remove, the less "stuff" that will interfere with the shearing process and get stuck in the fiber during processing.

- Place the alpacas in a smaller space, so they will feel more secure and you'll be able to quickly move them through the shearing process. This will mean less drama for you and much less stress on the alpacas.

- Buy three clear trash bags for each alpaca and prepare three labels for each. Mark them with the alpaca's name, year, and a 1, 2, or 3 for the first, second, and third portions of their fiber blanket.

- If you have higher-quality alpacas, have one small Ziplock bag ready to collect samples for fiber testing. You'll want these marked with each alpaca's name and ready to go. Higher-end alpacas may also require you to "noodle" the fiber blanket, so research this further if you want to have special sampling and testing done on alpaca fiber headed to a fiber show. We purchase large brown paper rolls to help with the noodling process.

- Have any immunizations and deworming shots prepped and labeled.

- Prepare a list of your alpacas with special notes for you and the shearers. We always have lots of pregnant ladies and at least one older female. We need this information front and center so the shearers know what to watch out for and who to be extra careful with while removing the fiber.

In addition to the above list, I also try to make sure we have a buddy system in place for younger alpacas or new arrivals. That way they can stay with their friends as they wait and then be back with their friends right after shearing is finished. In older, more established herds this wouldn't be needed.

Finally, we have a large area waiting for the alpacas after the shearing, so they can spread out, drink fresh water, and just relax. If it is a nice day, this will result in many alpacas sunbathing and taking naps in their new skinny and fiber-free bodies.

Extra Things to Know

A few final points you should know about shearing if you are a new or future alpaca owner in the United States:

- There are only a few shearing teams in the country, so make sure you research early and ask for referrals.

- It's hard to get on a team's schedule, so plan ahead and book your shearing months in advance. In the United States, we typically request shearing in January to get a spot on the list. The shearing schedules are published in March, and the teams arrive in April or May.

- Your shearing date is your date and changes are not allowed. These teams need to move quickly throughout the country, and they do not have the option of rescheduling.

- Our farm is always windy, which makes outdoor shearing difficult, so after our mistake the first year, we learned to set up inside the barn.

- If you opt for outside, make sure you have a tent set up to protect the shearing team from direct sun and heat.

- Biosecurity is super important, so select a shearing team that adheres to quality standards and best practices for cleaning clothing, boots, and equipment between farms.

- The cost of shearing is about $30 to $50 per alpaca plus a $100 setup fee. Be sure to budget for this, because it is a nonnegotiable item in any farm business plan.

- If you are a smaller farm, ask other local farms if you can shear at their location. This will save the setup fee and keep costs down.

- You'll need to have cash on hand to tip the shearing team. The tip scales along with your herd size and the amount of time the team spent at your farm. I usually tip 20 percent.

- The shearing teams work long hours, so feed them. I bake our team a homemade lasagna, have a place for them to sit and eat, then I send the leftovers with them. Since they are on the road for months, they appreciate a home-cooked meal.

Remember That Your Presence Will Make a Difference

One of our young alpacas, named Financy, is a very interactive alpaca who loves to spend time with me. I knew my presence would make a big difference for her on shearing day. I don't just get down on the mat to be next to her—I am literally in her face so she can focus on me, my smell, and my voice rather than on what the shearing team is doing.

When I had a C-section with my son Hunter, I had a reaction to the medicine they injected. I distinctly remember feeling my body react poorly, and I started to have a panic attack. The nurse bent down so she was very close to my face, and she spoke calmly and clearly while they injected another medicine to stop the reaction. I never forgot the nurse's actions because the interruption and authority she provided stopped my panic attack and gave me an immediate sense of well-being. This is the reason I do the same thing with my alpacas. If one of my herd is upset, I don't care what else is going on—I'm getting in their face so they can see, hear, and smell me. And when I do, you can instantly see their stress level decrease.

ALWAYS STAY CALM

Remind yourself to stay calm and relaxed on shearing day. Alpacas are emotionally intuitive animals, and they can sense your anxiety and stress. The more relaxed you are, the more secure your alpacas will feel.

I'm absolutely filthy after shearing, but it is worth it because I know my presence helps keep my alpacas calm.

CHAPTER 13

Maximize Your Profits from Fiber

What should you do with your fiber once it's sheared? It's a common question among alpaca owners, and there is no single answer. Every owner has a different situation, alpaca herd, and individual skill set, which creates the need for flexibility when determining what to do with the annual fiber harvest.

Visitors to our farm and shop often assume all of the products we offer are made from our alpacas and believe I personally make the items. Regrettably, both are false assumptions. I have zero crafting ability, and I am 100 percent incapable of knitting, crocheting, or felting.

But guess what? We still have beautiful alpaca products, and many do come from our own animals. Our farm is also profitable; we make good money from raising alpacas, having alpaca products made, and selling products across the country. We are successful because we rely on the expertise of others in our industry and community. I start with this because I want you to know you can make money from your alpacas and fiber, but to do so you need to understand what quality of fiber you have, what part of the processing you can do yourself, and what processing steps need to be outsourced.

The wall of yarn in our store. All yarn is grown in Michigan and made at cottage mills.

A Review of Basic Fiber Information

Firsts, Seconds, and Thirds

In the prior chapter on shearing, I mentioned segregating your sheared fiber into groups of firsts, seconds, and thirds. Before we dig into what you can do with these sections of fiber, let's review where they come from on the animal.

- **Firsts:** Also called the prime blanket, firsts come from the alpaca's back and top line.

- **Seconds:** Seconds come from the alpaca's neck area. Some farms will also use fiber from around the tail (our farm does not do this).

- **Thirds:** Thirds include fiber from the upper legs and belly.

Some alpaca owners will use every bit of the fiber and convert it into something usable. Others will retain only the firsts and throw out the remainder of the fiber because they just don't have a use for it.

Sections of Alpaca Fiber

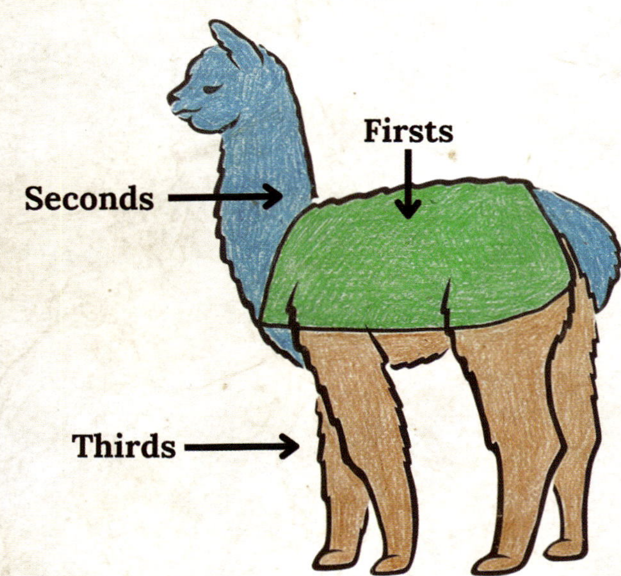

Alpaca Grades

Alpaca fiber is graded into classifications based on the fineness. While alpaca owners may know the fiber fineness based on sample testing for histograms or the EPD program, fiber mills and show judges generally grade fiber by sight and feel.

Grading fiber is a trained skill with professional courses that mill operators and judges take to become knowledgeable in this process. Alpaca owners can also take classes to learn this skill, and if you are interested in making the most money from your fiber, this would be a good educational opportunity to explore.

Fiber grades vary by geographical area in both verbiage used and micron count. For example, in the United States we use the phrase "royal alpaca" to refer to grade one, while in Peru they would use the phrase "superbaby alpaca" or "premium baby alpaca." The following chart provides a list of the grades and terminology used within the United States.

Fiber Grade	Classification	Micron	Example Usage
Grade 0	Ultra Royal	<16.9	Scarves, throws
Grade 1	Royal Alpaca	17–19.9	Scarves, sweaters, shawls, throws
Grade 2	Baby Alpaca	20–22.9	Hats, scarves, sweaters, ponchos, capes, shawls, throws
Grade 3	Superfine	23–25.9	Hats, mittens, gloves, socks
Grade 4		26–28.9	Socks, rugs, dryer balls
Grade 5		29>	Rugs, dryer balls

Knowing what grade of fiber you have will help you know what you can do with that fiber. I've given a basic list in the table; however, there are so many more uses!

Fiber Length

Another consideration is the length of your fiber. Wool fibers are made into goods via either a woolen or worsted processing method. Woolen processing can use short fiber lengths of 1.5 to 3 inches and worsted processing requires fiber 3 to 6 inches in length. For the most part, your fiber will be used for worsted processing. There are plenty of options for the fiber that doesn't quite make that 3-inch minimum, though, so don't let that minimum length concern you. Shorter fiber will reduce your processing options, but it doesn't stop you from using the fiber you've harvested.

Turning Fiber into Products

Now that we've covered fiber sections, grades, and lengths, let's review the large number of products that can be made from alpaca fiber. I have included fur products below, but keep in mind those products use a hide and can only come from a deceased alpaca

Firsts	Seconds	Thirds	Fur Hides
Hats	Rug yarn	Dryer balls	Stuffed animals
Headbands	Shoe insoles	Nesting balls	Hot water bottle covers
Mittens and gloves	Dryer balls	Wool pellets	Pillows
Scarves	Bird nesting balls	Stuffing for duvets	Slippers
Sweaters	Rugs	Stuffing for sleeping bags	Fur hats
Wraps and shawls	Trivets	Stuffing for dog beds	Fur earmuffs
Ponchos	Coasters		Fur blankets
Capes	Chair cushions		Fur rugs
Coats	Felt purses		Fur purses
Socks	Felt cat toys		
Throws and blankets	Baskets		
Yarn	Horse saddle pads		
Roving	Felted soap		
Felt	Felt		

Some of these products you can easily make on your farm, while others you would need a mill or commercial manufacturer to make for you. Now that we know what can be made, let's break out where you might have some of these products made. While this list isn't absolute, it will give you a general idea of manufacturing options available within the alpaca community.

Handmade by Alpaca Owners or Artisans	Machine-Made at Cottage/Micro Mills	Commercially Made in the US	Made in Peru via Artisans or Commercial Mills
Hats	Rug yarn	Socks	Hats
Headbands	Shoe insoles	Hats	Headbands
Mittens and gloves	Yarn	Scarves	Mittens and gloves
Scarves	Roving	Yarn	Scarves
Sweaters	Rug Yarn		Sweaters
Shawls	Felt		Shawls
Ponchos	Dryer balls		Ponchos
Yarn	Rugs		Yarn
Roving	Trivets		Stuffed animals
Dryer balls	Coasters		Pillows
Nesting balls	Chair cushions		Slippers
Felted Soap	Horse saddle pads		Fur hats
Baskets	Wool pellets		Fur earmuffs
			Fur blankets
			Fur rugs
			Fur purses
			Capes
			Throws
			Blankets

We use a hybrid approach to manufacturing on our farm. I can easily make nesting balls myself, but that's about it. We've had yarn, felt, roving (long ropes of fiber), rugs, trivets, shoe liners, and horse pads made in small batches for us at cottage mills, while we've run our socks through a commercial mill process. Commercial processing of alpaca fiber is very limited in the United States, but a few businesses have these capabilities and will accommodate large production runs.

Fur items are available through US wholesalers, or you can work directly with sources within Peru. For those just starting out, I would recommend using US wholesalers, as this is simpler and alleviates the need for international wire transfers and import declarations

And finally, I have a network of artisans who will turn our yarn into hats, headbands, mittens, scarves, shawls, rugs, and dryer balls for us. My mother-in-law, Lil, makes us hats, headbands, and ornaments. She has all the crafty skills that apparently skipped over me. Everyone always wants a "Grandma Lil hat," and they sell out very quickly.

The important takeaway here is you have lots of options. If you want to convert your fiber into products, you can do so yourself, you can use local artisans, or you can use many of the cottage mills located throughout the country.

Breaking Down the Fiber to Product Process

When we first started our farm, I didn't know anything about alpaca fiber, yarn, or making products. I remember wanting to cry at one point because I just didn't understand the process and the terminology. Since I'm not a crafty person, I found the entire concept completely overwhelming. Move forward a few years—I now know a lot about the process, and other farms come to me for advice. Who would have ever thought?

I don't want you to experience that same feeling of being overwhelmed, so I'd like to break down the process. This will help build a solid understanding about what happens to your fiber once it leaves your farm.

Here are the typical steps for processing raw alpaca fiber into yarn:

- **Shearing:** The fiber is sheared off the alpaca annually and divided into three groups called firsts, seconds, and thirds. We've covered this already, so you should be a pro on this part now. You're already farther along than I was at my first shearing!

Raw alpaca fiber starting its journey into yarn at Alpaca Fleece Mill in West Branch, Michigan. PHOTO CREDIT: FAITH RIEGLE

- **Skirting and sorting:** Once the fiber is segregated, we focus solely on the firsts, or prime blanket. Any heavy debris, like straw or hay, is removed and the fiber is sorted into groups by color, grade, and length. At this point, finer debris can be removed by "opening" or "tumbling" the fiber. This is done by hand or via a machine at the mill.

- **Washing or scouring:** The fiber next moves into washing, where a mild detergent is used to remove dust, grease, and vegetation. The fiber is then rinsed to remove the detergent and loose debris. Depending on the mill, this washing and rinsing process is repeated one or two more times. The mill will then carefully dry the fiber.

- **Carding and blending:** Once the fiber is fully dry, any excess debris will be removed. The fiber is next straightened and vertically aligned. Short, stray fibers are removed because they will degrade the softness and durability of the final yarn and products. The final process will blend the straightened fiber together into long fiber ropes called roving.

Alpaca roving at Alpaca Fleece Mill in West Branch, Michigan. PHOTO CREDIT: FAITH RIEGLE

- **Spinning, plying, and winding:** The roving is next fed into a machine to convert the large, long ropes into thin, single strands of yarn. The individual strands of yarn are then combined into two, three, or more strands to create a ply (such as 2-ply, 3-ply, etc.). Finally, the plied yarn is placed into a winding machine to create a yarn cone, skein, cake, or ball. Yarn cones will be used for machine knitting, whereas skeins, cakes, and balls are generally used for hand knitting.

At this point, the cottage mill would return the yarn to the alpaca owner to be sold to consumers as is or further converted into products such as hats, mittens, scarves, and shawls.

Inside a US Cottage Mill

The first year we had yarn made, we did so for our entire herd. I was caught off guard by the invoice, because I didn't understand the amount of processing that goes into the fiber. At that point I still had no idea what happened once the fiber left my farm.

Yarn on the production machine at Alpaca Fleece Mill in West Branch, Michigan. PHOTO CREDIT: FAITH RIEGLE

Many years down the road, I now have a friend named Faith Riegle who owns a cottage mill called Alpaca Fleece Mill in West Branch, Michigan. Thanks to her, I now not only understand the process, but I've been able to see the labor and expertise that Faith and mills like hers put into the processing.

Faith is newer to the industry, and I first met her when she took a tour at our farm. She came back many times to learn about what we do and what we want in our end-use products. I'm a stickler for quality and customer service, so Faith knew if she could please me, she could please anyone!

I applaud Faith for the research she has done and the investment she has made for our industry. All of Faith's equipment originates from Italy, and she has spent so much time learning her craft and fine-tuning it for alpaca fiber. I love talking to her and hearing about the evolution of her business.

I have visited the mill, and she has personally walked me through each step in the fiber to yarn process. While very similar to the process flow I provided above, it is more involved since she caters to smaller farms.

Following is Faith's twelve-step process for transforming raw alpaca fiber into yarn:

1. Incoming inspection for moths, mold, and tensile strength

2. Hand skirting to remove debris

3. Tumbling to remove additional debris, sand, and dirt

4. Hand washing

5. Flat drying

6. Picking and opening

7. Dehairing if needed (this removes longer, straight fiber)

8. Carding into slivers

9. Pin drafting into roving

10. Spinning into individual yarn plies

11. Plying into multiple yarn plies

12. Winding the yarn plies into cones, skeins, balls, etc.

Alpaca yarn cones ready for shipping at Alpaca Fleece Mill in West Branch, Michigan. PHOTO CREDIT: FAITH RIEGLE

Faith assesses quality each time the fiber passes to the next stage in her production process. If something starts to fail, she contacts the fiber owner to personally discuss and explore options. This is a level of service you wouldn't find in many other industries or at larger mills. Because Faith is operating a small cottage mill, she knows each customer provides fiber that is from an animal they love and that took an entire year to grow. That fiber has both emotional and financial importance, so she wants to make sure what she produces is perfect.

Making Commercial Alpaca Products

Commercial manufacturing is very different from the scenario I described above with Faith. Instead of creating a few skeins of yarn from individual alpacas, you're pooling your fiber harvest together with other alpaca owners so you can accumulate hundreds of pounds of fiber to move through the commercial production process.

We have been fortunate enough to become friends with Stacie Chavez at Imperial Yarn. Stacie is well known for her high-quality, American-made wool products. Stacie not only taught Jason and me the ins and outs of commercial processing, but also facilitates this for us through her company *and* shepherds our fiber from harvest all the way through to the end-use product.

Everything is done at scale. Instead of having a few large boxes of yarn skeins arrive, we now have a semitruck deliver pallets of American-made socks or blankets. I love the entire concept of raising alpacas and turning their output into warm, useful products for consumers. For me it always feels like Christmas when production completes and products arrive at the farm.

Pooling fiber with other farms requires a large cash outlay from us on the front end, but it is worth it. We get to have high-quality, American-grown and -made products for our store and we also help industry friends move excess fiber out of their own barns. It's good for us, good for the farms we work with, and good for the industry as a whole.

Alternative Fiber Uses

If you'd prefer to skip the process and cost of making yarn or products from your fiber, you can always sell the raw fiber to hand spinners. You'll find these folks at craft shows, via online stores, through marketplaces like Etsy, or in buy/sell groups on Facebook.

In the United States we also have a few fiber co-ops that will purchase your raw fiber and provide you with credits. You can then convert your credits into a check that you can cash or into alpaca products that you can use or sell.

The Future Is Fiber

For a large part of the 1900s, America was a major player in apparel production. That dominance began to slip away in the 1990s when machinery and clothing production moved overseas.

The United States now faces a shortage of machinery, expertise, and businesses ready and able to produce locally grown and manufactured apparel. The alpaca industry would love to change this, and while we don't have all the answers yet, we know our future is fiber and alpaca products can make a difference.

Each hat, scarf, and pair of socks we create helps bring clothing production back home. There is a growing awareness and distaste for fast fashion, fueling a movement toward products made from natural fibers like alpaca and merino wool.

No matter what you decide to do with your fiber, remember you are part of a community that loves their livestock and wants to do good, with individual members who are willing to collaborate to make alpaca fiber and goods mainstream.

CHAPTER 14

Setting Up a Farm Store and Agritourism

We started our farm with the idea of having a few livestock animals for entertainment and maybe a little side income. When COVID-19 hit in 2020, the alpacas became therapy and were key to maintaining balance and mental health. Michigan was hit hard in that first bit of Covid, our state was in lockdown, and I was secluded on our farm because I was immunocompromised. Whenever my stress would rise, I would head out to the paddock and sit on the ground with my alpacas.

As the pandemic continued, we started receiving calls from random people who were staying at Airbnbs in our area. Most were from Chicago and looking for something safe to do outside. As a goodwill gesture, we allowed people to come to the farm because we knew they were feeling as confined and restless as we were.

Then an interesting thing happened. We were told over and over again that those mini-visits were the best part of their trip to northern Michigan. We were also told we needed a store so people could support our business and thank us for allowing them into our home. After a few months of this, my husband and I discussed the possibility of building another barn to house a store. We could see the joy our alpacas brought to others, and we wanted to make this happiness available to more and more people.

A year later, we opened our new store and started hosting scheduled tours for the public. We had a grand opening weekend, the local news covered the event, and lots of people came out to experience the farm and shop in our store. It was an amazing weekend. We made a nice profit, but more importantly we made a lot of people happy when the world really needed happiness.

Our alpacas absolutely loved all the attention and would line up at the fence when they saw cars pull into the farm. They learned to pose for photos, entertained everyone, were gentle with small children, and embraced the wild activity with endless excitement.

When my husband would give tours, I would stand in our store and stare out the window in amazement. Each giggle and laugh told me we were making a difference and helping make people's lives better. And by we, I mean my quirky, loving alpacas.

During those first few years we had so many memorable guests. One woman drove hours to visit us before she started her chemo treatments, while another woman took a tour and came in crying because she was so thankful for the emotional release the alpacas provided after a difficult divorce.

Then one day we had a guest who stayed on our farm and hand-drew a picture of our cria Miss May that was absolutely stunning. She brought it back days after her stay and delivered it to my front door. It sits on my fireplace mantel today and will always reside there because it is a reminder of the good alpacas can bring others.

Little did I know my beloved alpacas possessed a magical ability to connect with humans and make them feel

My gift from a Harvest Host guest.

loved, special, and at peace. All these little micro-moments showed us and our community that agritourism was a powerful force taking over America.

The Rise of Agritourism

Agritourism is a newer term and not everyone is familiar with it. The US Department of Agriculture defines it as "a form of commercial enterprise that links agricultural production and/or processing with tourism to attract visitors onto a farm, ranch, or other agricultural business for the purposes of entertaining or educating the visitors while generating income for the farm, ranch, or business owner." Before we created an on-site store and opened for tours, we visited our local planning board to teach them what agritourism was and explain how it would benefit us and the community. The board loved the idea and gave us "kids" their blessing to move forward.

As we jumped in, we did so with full gusto! We built a 1,200-square-foot store, which is much larger than the average farm store. We stocked it with hundreds of alpaca products, which again was a bit overkill. We set up formal interactive tours, welcomed tour buses, hosted homeschool groups, and ran birthday parties and day camps. Each year we would have between five thousand and ten thousand visitors come through the farm. We met people from all over the world and welcomed return visitors year after year.

Display options inside our farm store.

Opening an On-Site Farm Store

There is no better way to introduce consumers to alpaca products than by having them available in an on-site farm store. Not only does it help people connect apparel, home goods, and toys to the source, it also allows visitors to take a bit of the farm home with them.

Before we opened our store, I researched a lot about retail and tried to implement best practices. But information in books and blog posts just doesn't compare to the learning opportunities available from being in the store and interacting with customers. Not only could I watch customer movements and their interaction with products, but I could also listen to their discussions and answer their questions. All of this provided invaluable lessons that helped me improve our store operations, better serve customers, and become more profitable. I'd like to share my lessons learned with you, because small changes can have a big impact on your success.

- **Offer items at a variety of price points.** People who visit will have vastly different budgets, so you need to offer lower-cost items like pens and magnets as well as higher-cost items like sweaters, coats, and blankets. This will let you please everyone and make sure you don't miss sales.

- **Offer items that cater to both genders and all ages.** Women love sweaters, scarves, and throws, while men love socks and hats. Kids love stuffed animals and little items that keep their hands busy. Children's clothing rarely sells, and honestly, the kids hate it, so don't spend a lot of time and money on these items. Children want toys, so stock up on the stuffed animals and toys, then watch those little fluffy alpacas fly off the shelves.

- **Know what your area can bear for pricing and don't push beyond what is feasible for your area or visitors.** We are just outside a pricey tourist area with a very nice ski resort close by. This allows us to sell goods at a traditional markup, while other farms in lower economic areas must sell their products at a much lesser profit margin.

- **Have something for all seasons.** Trivets, nesting balls, dryer balls, and lightweight socks are great options for summer sales. People will buy fewer winter goods in the summer heat, so make sure you have household items in warmer months to encourage purchases.

- **If items are made from your animals, make sure you tell people.** Stating exactly what animal a product came from helps shoppers emotionally connect to the product. This

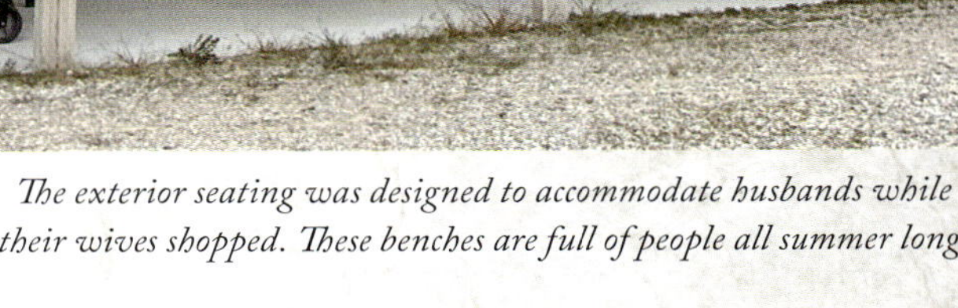

The exterior seating was designed to accommodate husbands while their wives shopped. These benches are full of people all summer long.

Dryer balls make the perfect item for summer sales. They also make great impulse items.

increases sales. You can simply write the alpaca's name on a product tag or you could have a photo of the alpaca close by.

- **Have photographs of your alpacas on your store walls.** Seeing pictures of the alpacas inside the store will help visitors remember this is a farm and your animals made the products you're selling. It will make the buyer feel good about their purchase!

- **Put small, impulse items by your checkout.** These could be little magnets, dryer balls, or tiny stuffed animals. They will be added to the existing purchase, which increases the average order value and your revenue.

- **Clearly state where goods are made.** If you make something in the United States from your animals, make sure this information is apparent. If something comes from Peru or another country, document this to help people know it isn't from your farm.

- **Ditch the plastic.** Each time I see a lot of alpaca store owners leaving plastic on products, it makes me sad because I know it is costing them sales. Alpaca products are luxury items that need to be touched and felt. You cannot do this through plastic, so make sure you remove all plastic before items go into the store.

- **Skip food and dated items like calendars.** Food and date-specific items won't sell in the time frame needed, so you'll end up throwing these out. This will reduce your overall profits.

K   Katie
14 reviews · 0 photos

★★★★★   Jun 9, 2022

Hands down the BEST experience -- if you're even considering booking a tour of the farm or visiting the store, please do!! Even as an adult, I had so much fun meeting, petting, and feeding the alpacas. They are such a delight. The store is also wonderful - I got a lovely winter hat made from the fiber of Dolly, who I got to take a picture with at the barn! Rebecca and Jason are so friendly, accommodating, professional and great to talk to. They're wanting to make sure everyone, from littles to adults, can walk away with great memories - and they do a top-notch job at that. Do visit if you're in the area!

Notice the review specifically calls out Dolly and the hat made from her fiber.

- **Use a good point-of-sale software solution to help track inventory and sales.** We use Square's retail software, and it has been well worth the cost. I know exactly what inventory I have in stock, all my inventory costs, and what days and months are most profitable, and I can track top product sellers.

- **Do not charge extra for credit card transactions.** Alpaca products are already at a premium and you don't want credit card surcharges to hinder a sale. These fees should be included in your operating costs and covered within your product markup.

- **Showcase your store on your website.** People will look at your website before they arrive, and if they don't like what they see, they may skip the visit entirely. If you don't have an online storefront available, include photos of your store so farm visitors can see what you offer, which will encourage them to come and shop.

Agritourism Options for Alpaca Owners

The beautiful part of agritourism is that you get to define what you want to offer and how big you'd like to scale. You can open your farm occasionally, monthly, weekly, or daily. You can offer anything from tours and photo shoots to birthday parties and sip-and-shop events. Because it is your farm and your business, you decide what works best for you.

We've taken this approach to heart, and we've tried a lot of different aspects of agritourism. During the process, we realized not everything is right for every farm. There were aspects of agritourism that we loved and aspects that weren't for us. As we tried different revenue streams, we considered the profitability as well as the happiness factor it brought to our family. Along the way, we realized there were things that weren't worth the time, money, and energy, so we adapted. This introspection and openness to pivot has kept our farm profitable and our family at peace.

As you consider adding agritourism activities to your farm, I highly encourage you to do the same kind of evaluation along the way. Explore different options and zero in on the ones that satisfy both your business goals and your life goals. Say goodbye to the ones that do not make you happy—and know this isn't a failure; instead, it is a honing-in on what works. This took my type-A personality awhile to accept, but once I did, I was a lot happier.

Let's explore some of the agritourism options available and review some pros and cons associated with them.

Interactive Tours and Trekking

Interactive tours produce great revenue while requiring very little investment. Tours are appealing to all ages, from small toddlers to senior citizens. Because alpacas are friendly and docile, it is safe to have humans inside the paddocks with them.
Some recommendations that will help make tours successful include:

- Use a tour scheduling software to automate the ticket sales process. People want to book tours at all hours of the day and night, so make it easy for them.

- Have staggered pricing with different price points for adults and children. We have kept children at $5 per ticket so as many children could come as possible. We set adults at $10 a ticket because we realized we have lots of adults coming back time and time again, so this price point is considered economical.

- Don't price your tour tickets any higher than your local movie ticket prices. This keeps your tours competitive.

- Overcommunicate and set expectations. The more information I provide at the time of purchase and via reminder emails, the fewer calls we receive on tour days.

- Provide a small bag of food for each person to use on their tour. This teaches your alpacas that humans are fun and they bring treats. But on the flip side, don't provide too much food! This will make your alpacas overweight and quickly degrade their fiber quality.

Private Group Tours

Once you set up your tours, you'll start receiving requests for private group tours. These could be family reunions, homeschool groups, day care centers, or senior centers. Your time is valuable, so charge appropriately for these tours.

Tour Buses

Tour buses are a fun and great way to get people onto the farm. We charge for the tour bus and then have the store open for shopping before, during, and after the tour. This

will allow you to meet some fun folks and sell quite
a bit of goods in a short period of time.

Contact your local tourism board to locate tour
operators that visit your area. You might be surprised at how many come through and how easy it is
to get on their schedules.

Birthday Parties

Kids love alpacas and moms love to get their kids
outside in nature, which makes alpaca birthday parties a fan favorite. We have handled these like group
tours but with an optional add-on fee for gift bags.
Many moms have been happy to take us up on this
offer, which helps sell small alpaca products. Some
parents also visit the store and shop during or after
the party.

Make sure you have an ample supply of hand wipes and garbage bags available. We discovered moms bring frosted cupcakes, but often forget anything to clean up those sticky little hands before they go inside to shop. Colored frosting does not mix well with high-end alpaca products!

Day Camps

There are a number of alpaca owners offering summer day camps. We ran these one summer and held multiple-day camp sessions for two different age groups. I'm really bad at crafts and entertaining kids, so we hired one of our 4-H graduates to take on this task. She did an amazing job and the kids had lots of fun.

If you decide to offer these camps, plan early, promote them on social media, and have a way for moms to book online. Parents plan for these camps well in advance, so you have to be early in your planning and scheduling.

Photo Shoots

Photo shoots can be offered to individuals, families, or professional photographers. We've had photographers use our farm for specialized shoots with families

PHOTO CREDIT: DAVE SPECKMAN FROM SPECKMAN PHOTOGRAPHY, LLC AND NORTHERN BLESSINGS ALPACAS

in exchange for some wonderful professional photos of the farm and animals. Our friend Dave Speckman is a professional photographer and alpaca owner at Northern Blessings Alpacas. They offer the most adorable mini-shoots at Christmastime for families and couples.

Harvest Host

The Harvest Host program is a great option for attracting travelers to your farm. After joining the program, you set a calendar of availability, guests will request visits online, you approve the visits you can accommodate, and the guests arrive with their RV or camper to stay one night on your property. In exchange for the stay, the guest will spend money on your farm by purchasing products or services. Harvest Host loves alpacas and alpacas love Harvest Host guests. It is free for farms to join the Harvest Host program, which in turn brings guests to your farm and helps promote your farm to members throughout the country. The first year we participated, our average Harvest Host spend was well over $200 per visit. It's a great program that offers a lot of perks to both hosts and guests.

Weddings

Over the years we've taken our boy alpacas to several weddings. The wedding couples love it, and the guests find it a great surprise. The alpacas mingle with the guests during cocktail hour, which allows guests to hand-feed them and take photos.

Some recommendations for making the event flawless:

- Have formal contracts so the bridal couple knows exactly what they can expect at the wedding. This is a big day for them and they'll want you to over-communicate.

details. Brides will inquire about dressing the alpacas in costumes or flowers, having the alpacas carry baskets of wedding favors, or options for photos. Knowing what your alpacas can handle and communicating this to the bride and groom will help make the event a positive experience for everyone involved.

- Touch base with the wedding venue ahead of time so you know where to park, what to bring, and if there are any restrictions.

- Bring equipment to clean up after the alpacas. You will need it.

- Keep a close eye on your alpacas. At one wedding, Chase was huddled up with the bridal party taking pictures and we suddenly realized he was drinking a bridesmaid's gin and tonic! The rest of the evening he was hyper-focused on the plastic cups and trying to sneak sips from anyone and everyone.

Getting People to Your Farm

One of the most challenging things about holding agritourism activities at your farm is actually getting people to find out about your offerings and come visit. Once you get the momentum going, visits will flow without a lot of work. That said, you have to get the ball rolling first. Here are my tips for gaining exposure:

- If the media calls, say yes. Bring them onto your farm, take alpacas to their studio, and do anything else you can to get that free exposure. It is intimidating, but well worth it.

- Contact your local tourism group to see what programs they offer. These groups are always looking for something new and exciting to offer visitors.

- Create rack cards and put them in hotels, stores, and restaurants. This is a low-cost, great way to get your name out.

- Use social media platforms like Facebook, Instagram, or TikTok. You don't need to be a professional to grab attention and gain followers.

- Partner with local wineries, distilleries, ski resorts, or other businesses. Many of the farms in our area will take their alpacas to local businesses for family nights or special weekend events. This is a great way to build your brand, and you can make money while doing it.

When we hosted our first open farm day, I was hoping for a few hundred people. We had nonstop visitors all day long—about fifteen hundred people came in one day! This taught me a very valuable lesson. I don't need to be a professional in marketing to get people to the farm; I just need to let the alpacas be their cute, quirky selves. They bring the humans and they keep the humans coming because no one can say no to those adorable and funny faces.

CHAPTER 15

INVOLVING THE YOUTH

When we decided to venture into the alpaca lifestyle, we had no idea there was a rich world of youth activities waiting for us. We knew alpacas would be good with children, but we didn't realize the industry was set up to support youth through 4-H and sanctioned shows of the Alpaca Owners Association (AOA). Getting youth involved in alpaca-related activities can provide a physically and emotionally safe environment. Showing alpacas helps young people learn about the industry, builds confidence, and provides a foundation for future business opportunities should they decide to take this route.

	AOA	**4-H**
Membership required	Youth	Youth
Location of shows	National and regional	National and regional
Classes available	Showmanship Public relations Costume Obstacle	Showmanship Public relations Costume Obstacle
Attend educational events	Not required	Required
Attend volunteer events	Not required	Required
Scholarship opportunities	No	Yes
Cost	Variable	Variable

I'd like to share our experiences with both of these organizations so you can see the massive impact they can have on participating youth.

Alpaca Owners Association Shows

We took Hunter to his first alpaca show in 2019. We started with the Michigan International Alpaca Fest, which is an in-state, primarily adult event. We did not bring our own alpacas; we simply went to see, learn, and experience the event.

Before I realized it, my son was helping out other alpaca owners and sucking up every bit of knowledge his adolescent head could absorb. Every adult we encountered embraced and welcomed him, patiently answered his questions, and went out of their way to provide education. Hunter was in heaven. When I look back at the photos from that day, I can see happiness all over his face.

Hunter with Heartthrob at Nationals.

The Michigan show left a lasting impression, and Hunter's excitement was infectious. We realized we needed to give the show system a chance, so we purchased two young ladies named Heartthrob and Onyx. Hunter, Jason, and I were all getting ready for the National Alpaca Show in Indiana when Covid hit. At that time no one really understood the impact it was going to have, so the show went on. I was in the middle of packing when we started to hear of farms canceling their plans to attend. Since I'm immunocompromised I stayed home, but I felt comfortable still sending the boys. While our alpacas didn't win anything beyond a third-place ribbon, Hunter fell deeper in love with the show system. He liked the ribbons, but you could tell the adventure and the time he spent with adults was what really captured his attention. He came home with a fancy association T-shirt and an educational book on showing alpacas. Both were gifted by our future mentors, Kim and Nancy. He was so excited to tell me everything about the show, how Onyx and Heartthrob did in the show ring, and *everything* he learned from Kim and Nancy. As a parent, all I wanted to do was nurture that passion and enthusiasm.

I have zero interest in showing or the show system, but the men in my life did, and it became a family event for us. For the next few years, we worked to upgrade our herd. Hunter and his dad memorized lineages for all the alpacas they found interesting. They would sit together and pore over data to discuss top genetic lines, fiber characteristics, and show results. I did my best to look interested and listen as they both rambled on and on about alpacas.

For our family, showing alpacas was quality time spent together. Instead of heading to Disney, we would head to an alpaca show. We didn't show a lot, but when we did it was memorable. Hunter is now at college, and I realize how much I miss our long conversations about alpacas and our time together at the shows.

If you'd like to experience the same, there are many ways to get your child involved in the show system. Many AOA shows have special competitions that offer the opportunity for children to get into the ring and participate. These "classes" include Youth Showmanship, Youth Obstacle, Youth Public Relations, and Youth Costume. Children from ages five through eighteen can learn about handling alpacas, safety, animal husbandry, and communication.

The beautiful thing about AOA shows is that your child can decide to participate in youth competitions, adult competitions, or a combination of both. And their participation can vary by show. This provides a lot of opportunities for children to learn about alpacas, gain skills, and become further involved in the alpaca industry.

AOA shows are a great way to get youth involved in the alpaca industry, teach them about community, and allow them to mature through the showmanship experience. If you're interested, remember that you don't need to have an alpaca to get started. You can just show up to watch and learn. And if you're lucky, a mentor farm will find you and work to get you involved. I am so thankful to our mentors, the time they've spent with my son, and the life lessons they helped teach him along the way.

4-H Fairs and Events

The 4-H is the largest youth development organization in the United States, and it is a powerful force in getting children outside and involved with animals. With over one hundred public universities across the country involved, the 4-H program has provided

4-H stalls and banners at the Northwestern Michigan Fair.

life-changing experiences for more than six million children. It is a powerful and impactful organization that I'm proud to be part of.

Shortly after starting our farm, our family became involved with the local 4-H program. Before I knew it, my son and his trusty alpaca Adel were herding humans into the alpaca barn, providing education, taking photos, and interacting in ways I didn't think were possible. My introverted son blossomed, and when Adel was by his side, he was unstoppable.

At some point I blinked and my husband was now in charge of his own 4-H alpaca group called Farm and Fleece. We started to "rent" our alpacas to local children so they could participate in the 4-H fair. In exchange for using our alpacas,

Hunter and Adel at the fair.

Hunter's smile sums up 4-H fair week. It offers so many memories for everyone in the family.

Hunter and I at his last 4-H fair. It was bitter-sweet to have this all come to an end. I'm so proud of his growth, but I will deeply miss the family time.

the kids would come to our farm to work off their debt. This "work"—mostly training the alpacas and helping out at farm open houses—was really just code for getting the kids on our farm and out in nature.

Each year my husband and son would disappear for a week of camping and show activities at our local livestock fair. They would come home with stories, photos, and sleep deprivation. Then a week later they would already be talking about next year's fair, what alpacas we should take, and what we could do to improve the program for our group.

As an added bonus, each year at the fair I get to see alpacas I've previously owned. We've ended up selling over twenty alpacas to 4-H families, and they fill a good portion of the camelid barn. For me, this is the frosting on an already-amazing cupcake. I'm smiling the entire time I watch the kids and alpacas compete, because I've been able to watch them both grow throughout the years they've been part of the 4-H program.

This last year we had a mix of first-time participants all the way up to seniors who came back to show.

The 4-H organization uses the phrase "4-H grows here," and it is so true. Here are a few standout moments from 2024 that show why 4-H matters:

- Our smallest Cloverbud, Allie, learned to handle an alpaca twice her size. This was Allie's first year and she was tiny, but mighty.

- Young Marlee returned with her alpaca, Seven. They have such a strong bond that Marlee can maneuver Seven through any obstacle course with a big smile on her face. Her excitement is infectious and her love for Seven is endless.

- Halle somehow managed to fit 4-H training in between multiple sports, and you can see the love she has for her alpaca year after year. After the fair, Halle's alpacas always sell quickly because of the work she has put in and the trust she has carefully fostered throughout her training sessions.

- Tessa trained daily with her alpaca, Ryan, only to find out he had a severe case of "ants in his pants" in every competition. He was so busy watching the crowd of people he forgot everything Tessa trained him on. As I watched from the sidelines I found myself getting frustrated with Ryan because I knew how much time Tessa had dedicated to training. But Tessa handled it with absolute grace and patience. And while they didn't win a banner as she had hoped, she grew from the experience and Ryan now runs to greet any human he sees.

- In his final year of 4-H, Hunter learned that a fair isn't just about winning a banner. As a senior, he was more focused on mentoring the younger children and helping them meet their individual goals. I saw him work with younger 4-H participants, cheer them on, and celebrate their wins—even when that win meant he didn't receive a banner or trophy himself. As all of this unfolded, I could not help but think my introverted child was ending his 4-H career at what I would consider the very top of excellence. Throughout the years, he moved from wanting to win to wanting to see others win. That shows the true power of 4-H and the impact it has on personal growth and maturity.

Marlee and Mr. Seven

Because we have a dedicated barn for llamas and alpacas, it becomes a home away from home for all the camelid-loving kids. The 4-H participants have schedules for cleaning, animal care, and interacting with the general public who stroll through looking at the animals. You rarely hear a child complain about their duties because they know the importance of what those duties entail.

I know not every barn has the comradery that our group does. We are blessed to have barn leaders who create an atmosphere where youth matters and who lead by example, take time to celebrate, and make sure our participating kids have a fair week they will never forget. Ben, Kathy, and Renee are the heart of our 4-H program, and every kid who comes through that barn is better for it.

Ben, Kathy, and Renee celebrating Hunter's banners at his senior year of fair.

Many rural areas don't have an existing 4-H program for alpacas, but that doesn't mean you cannot start one yourself! The national 4-H organization has a lot of resources to help communities get started. I promise you, the children that participate will be changed by the experience, and the program will be worth every minute you dedicate to it.

Ribbons and So Much More

Having the pleasure of watching many kids and their alpacas participate in competition has given me a great deal of appreciation for these groups and associations. The kids win ribbons, but there is so much more to the program and the week at fair.

A few years ago, we wrapped up day one of the 4-H competition and I was adamant about getting a photo of Tessa and her ribbons. There were a lot of kids that walked away with ribbons after doing a great job with their animals, but Tessa stole the show. She had spent months working with two alpacas named Amara and Escobar.

*Ryan kissing Tessa in the middle
of the obstacle competition.*

Tessa and her ribbons after day one of the fair.

Amara had a special place in Tessa's heart because Tessa was at her birth, and she saved Amara later that day when the cria started to have seizures. I was working in our store and Jason was doing a tour when Tessa came running out of the barn to get us. Tessa and Jason rushed Amara to the vet hospital at Michigan State University while I took over the tours and Tessa's mom, Jet, ran the store. Tessa held Amara in her lap the entire way down to MSU, and while there she decided on Amara's name. Amara pulled through and from that day on she was Tessa's girl.

Tessa wanted to use Amara for 4-H but we were hesitant. Amara is Nibbler's sister, and you could tell she shared some of her sister's anxiety issues. This didn't stop Tessa. Day after day, she worked to slowly build Amara's trust. I'd see them just sitting together on our farm, having chats like I always did with Nibbler when she was young. Once Tessa had gained Amara's full trust, she moved on to halter training and followed that with obstacles.

Tessa's dedication, perseverance, and true love for the alpaca paid off and she walked away with a ton of ribbons. But what really impressed me was how Amara showed very little stress throughout competition because she knew she was in the safe hands of her friend Tessa.

I was so impressed with this pair because I could see how much they grew over that summer. Tessa learned that hard work pays off, and Amara learned that all things are possible if you have a trusted human by your side.

Tessa and Amara illustrate why alpacas are so much more than fiber. They have amazing superpowers to connect with youth and to effortlessly teach them life lessons.

CHAPTER 16

SELLING ALPACAS

One of the most challenging aspects of owning and breeding alpacas is *selling* them. The reason isn't because the market lacks available buyers. When alpaca transactions are nonexistent, it is because the owners simply don't yet know how to optimize the sales process.

We've sold over one hundred alpacas in four years. If you add the number of alpacas we've sold for other farms, that number jumps much higher. And guess what? We generally only have about fifty alpacas on our farms at a time, we don't have the highest-quality alpacas, and we don't sell at the lowest prices. We do well because I've optimized the sales process so that it can work well for both us and the buyer. That last part is important. The key to success is making sure you focus on the buyer and their needs.

In this chapter, I'll walk you through our process and explain my secrets to success.

Deciding Who You Should Sell

The first step in any sales process is to figure out what you have for sale. That sounds easy enough, but you would be surprised at how many alpaca owners don't have a ready list of who is available for sale.

Fixing this problem is easier than you think. It starts with an impartial evaluation of your herd. We've been doing herd evaluations since we first started our farm. We do this to help decide which alpacas meet our goals and which alpacas might meet the goals of our customers.

A herd evaluation should consider the data points that matter to you, your personal goals, and your business goals. For us, these include things like histograms,

EPDs, lineage or pedigree, color genotype, show history, breeding ability, cria care, personality, phenotype or appearance, fleece weight, available fiber usage, age, and visible color.

The following chart illustrates how we would perform an evaluation. I've used actual alpacas on our farm to help illustrate the process. The key takeaway is that if you use data to reduce emotions, it makes it clear who should be for sale and also helps establish a solid pricing model.

SAMPLE ALPACA EVALUATION MATRIX

	Nibbler	Diva	Dolly	Maui Wowie
Lineage	Good	Good	Average	Excellent
Show Record		Good	Good	
EPDs			Top 10%	Top 1%
Fineness	Good	Average	Average	Average
Crimp	Good	Good	Average	Good
Staple Length	Excellent	Good	Excellent	Good
Density	Good	Good	Excellent	Good
Uniformity	Good	Good	Average	Good
Fleece Weight	Excellent	Good	Excellent	Good
Conformation	Good	Good	Average	Good
Phenotype/ Appearance	Good	Excellent	Excellent	Good
Color Phenotype	Brown	Dark Silver Grey	White	White
Color Genotype		EE a3a2	ee AA	ee Aa1
Ability to Breed	Good	Good	Good	Good
Ease of Birthing	Good	Good	Good	Good
Cria Care	Poor	Good	Excellent	Good
Offspring	Average	Good	Good	Excellent
Personality	Average	Good	Average	Good
Age	5	4	7	7

This matrix is great for objectively comparing one alpaca to another, but it doesn't tell the entire story. Now that we've looked at the data, we need to compare it to how we would use these ladies versus how a potential buyer might use them.

SAMPLE ALPACA USAGE MATRIX

	Nibbler	Diva	Dolly	Maui Wowie
Breeder for Show		X		X
Breeder for Fiber		X	X	X
Fiber Production	X	X	X	
Agritourism	X	X	X	X
4-H Participant				
Retired				
Nonnegotiable	X			
Sell Price		$7,500	$3,000	$5,000

Our farm has multiple revenue streams, and we need to own alpacas that can serve our business in a variety of ways. We use this matrix to guide us, but it is not the final factor in decisions.

Let's use Dolly as an example:

- While she did receive a Reserve Championship banner years ago, her crimp structure is bold, and she wouldn't do well in today's competitive show ring.

- We've discovered Dolly's bold crimp structure is a strong trait that passes to her offspring. Dolly tends to overpower a male with excellent crimp structure, and while the male can improve her, it is not enough to win at a show. Because of this, Dolly would be great for producing alpacas for fiber usage, but we wouldn't use or sell her to produce show stock.

- Dolly is pint size, which makes her conformation not ideal. She shouldn't be bred to large males because she will struggle during birthing.

- Her density and staple length are wonderful, which makes her a top fiber producer on our farm. You wouldn't think this would be the case since she is smaller in stature, but make no mistake, Dolly is a powerhouse of fiber production.

- Dolly is an excellent mother, she is wonderful with visitors, and her sweet and sassy personality makes me happy.

- Dolly was one of our original alpacas, which gives her a special place in my heart. I know this because whenever we talk about selling her I well up with tears.

So, the magical question is do we sell Dolly, or do we keep her? Dolly is for sale; however, it would take the right person for me to sell her. If someone came to our farm and they fell in love with Dolly—and I could see Dolly loved them—I would sell her. But because I love her as much as I do, it would have to be an excellent match. The end result is Dolly goes on the list of alpacas for sale. I know I cannot keep all my alpacas, so I have to prioritize who is for sale and who stays on the farm. This list, however, does change based on our farm needs and the needs of buyers.

Dolly

Now let's look at Diva:

- Diva is solid in every way, and she is beautiful inside and out.

- She comes from a great lineage and her first offspring was very nice.

- She produces lovely fiber that I eagerly make into hats that sell out quickly.

- She is an excellent mother and an overall easy keeper.

- She is young and would make an excellent dam for a show-focused farm.

Should we have Diva for sale? She is for sale, but only because she is no longer a great fit for our business goals. Now that our son is an adult, we rarely go to alpaca shows. Diva is a show girl, and she produces show offspring. Diva's genetics are good for the industry and we want her to be used, so she is for sale. Because Diva is excellent in every way, she is listed at a higher price.

Now let's talk about Nibbler:

Nibbler is my number one nonnegotiable.

- Nibbler is my girl and has been my girl since the day I met her at Fun in the Country Alpacas.

- We've had Nibbler since she was six months old. I have a strong bond with her, and I honestly cannot imagine having an alpaca farm without her part of the herd.

- She is a top fiber producer on our farm and produces enough fiber each year to make eighteen double-knit hats. That's a lot of fiber!

- She has anxiety issues, which impacts her ability to do many things, so she doesn't fit into many of our usage buckets. Her anxiety is so bad, she isn't fit for breeding or 4-H.

Should I sell Nibbler? Nope. It wouldn't be good for her, me, or a potential buyer. That girl will be the very last alpaca to leave our farm, and I truly hope I get to love on her until the day she dies. She is what we call a "nonnegotiable," and my husband knows once someone is classified as such they are not up for discussion.

Sometimes you'll have an alpaca that tugs on your heart but may not fit within your business plan. On our farm we call these "nonnegotiables." They are off limits and our family knows that alpaca is not leaving. I have about five of these alpacas, and my son, Hunter, has three. And that's okay. It allows us to breed and still be happy.

Hunter caring for his nonnegotiable alpaca Chase during a medical procedure.

Before we started using data and classifying alpacas into buckets, I had a very difficult time deciding who should be for sale and how they should be priced. I found the entire process emotionally draining. Using data and including a veto option has made things a lot easier.

Creating a Good Sales Page

If you plan on selling alpacas, you must set up a good sales page to market the animals. Many alpaca owners assume this isn't necessary because they believe they can simply email details, verbally share information, or physically show a buyer the alpaca when they visit. However, in doing so, prospective buyers might never know you have the perfect alpaca for their needs.

It is easy to get caught up in the daily activities of farm life and neglect your website or forget to market the alpacas you have for sale. When this happens, you aren't providing value to prospective buyers to reach out, and you won't know if you've missed a golden opportunity.

Crafting a solid sales page is a critical part of the sales process. Compiling information to put on this page requires time and patience, but it is worth doing.

Here's what you need to include on your sales page for each alpaca:

- A primary image

- Key details such as registration number, birthdate, sex, breeding status, color, sire, and dam

- Histogram data and PDF download (if available)

- EPD data (if available)

Diva

Pedigree

Diva offers fabulous black and grey champion genetics. She is sired by multi-champion, Captain Midnight, crossed with A Paca Fun's Intrigue on her dam's side. Captain Midnight has commanded ten Champion banners, seven Reserve Champion banners, and twenty-four first-place ribbons.

Phenotype

This girl has it all! She has a well-organized crimp structure that reaches into her underbelly, her conformation is perfect, and her fiber has a soft handle. Diva stands out in a crowd because she is beautiful. Her fiber color is a lovely silver grey that is nicely complemented by her striking black eyes and thick eyelashes.

Personality

Diva's name is a far representation of her personality. She is laid back, friendly, and she is eager to give you a kiss while flashing her beautiful dark eyelashes. She blends into the herd due to her calm nature, then brings out all the love when humans arrive.

Color Genotype: EE a3a2

Diva's ASIP alleles show she has a black allele in both locations and her MC1R alleles show she has no dilution. This genotype combination tells us she cannot produce light colored offspring. She will produce offspring that are dark fawn or darker, with a tendency for darker colors. Diva also covers the classic grey gene, which makes her a wonderful addition to any grey breeding program.

General Information

Highlights: Show Girl & EE a3a2
Type: Female Huacaya Alpaca
Status: Proven Breeder
Birthdate: 09/10/2020
Official Name: CSAF Drama's Silver Diva
Registration: ARI Registered
ARI No: 36086238
Color: Dark Silver Grey and White

Offspring

Males:

Cotton Creek's Enchanted Divo (DSG/WH)

Fiber Statistics

2022
AFD: 22.5
SD: 3.9
CV: 17.2
CF: 96.1%
Curve: 43
Staple: 100 mm

Lineage

Sire: Captain Midnight – ARI# 32401714 (TB)
Dam: A Paca Fun's Drama Queen – ARI# 32118582 (MSG)

Awards

- 1st – 2022 Buckeye Fall Fest Full Fleece Huacaya Grey Female Two-Year-Old
- 2nd – 2022 AOA National Show Fleece Huacaya Silver Grey Female Yearling
- 2nd – 2021 Green Mountain Alpaca Fall Spectacular Walking Fleece Huacaya Silver Grey Yearling
- 2nd – 2021 Green Mountain Alpaca Fall Spectacular Halter Full Fleece Huacaya Silver Grey Female Yearling
- 3rd – 2021 New England Coastal Classic Halter Full Fleece Huacaya Silver Grey Female Yearling

Documentation

Certificate of Registration

Histogram

Price: $7,500

Silver Diva's sales page is a good example of the information to offer potential buyers.

- A detailed description of lineage, phenotype, and personality

- Offspring (if available)

- Supplemental images that illustrate conformation, fiber blankets, and personality

- Price

A common practice (and mistake) in the world of alpacas is to use "call for price" in lieu of an actual price. Sellers do this for a couple of reasons. First, they may not know what price to set for the animal. We've solved this issue with our matrix and the data it contains. The second reason sellers do this is because they want to modify the sell price based on the buyer's budget. This one annoys me a little, and I simply don't believe in this approach. When we first started our farm, we were told "the price of the alpaca is what the buyer will pay." This statement goes against everything I believe in. I believe the price of the alpaca is based on the alpaca's available use, the current market conditions, and the seller's need to move animals. If you fail to set a price, you'll fail to get inquiries for your animals. It's that simple.

It is also important to create a description that is consistent with the animal's intended use. If I am selling an alpaca to be used for 4-H or therapy, the description should focus on those key personality traits, and the images should illustrate the animal interacting with humans. If I'm selling an alpaca as a breeder of show stock, then the description should primarily focus on show results, offspring show results, and fiber characteristics.

As you create your sales pages, just keep a tight focus on who you believe would be a good buyer for this alpaca and match the description to this buyer. This process is similar to staging a house. Help the future owners envision how this alpaca will fit their needs and goals. That's the secret to selling!

FLOODING FACEBOOK ISN'T A SALES STRATEGY

Many alpaca owners will rely solely on Facebook groups to sell alpacas. While Facebook is a good marketing tool, it is not a strategy. It is one way to reach buyers, but it is not the only way. A large percentage of buyers won't use Facebook, so don't rely on this as your only sales channel.

Teddy and Levi were sold to a family in South Carolina when they were about a year old. The new owner joked that he had to purchase alpacas from states away because his daughter was dead-set on these boys who were "best friends." Good marketing lets you sell well outside your area.

Qualifying Interested Buyers

You've decided who to sell and you've successfully marketed your alpacas so that you have an active prospect. Congratulations! Now it's time to qualify the future buyers to make sure they are a suitable home for your alpacas.

Some important questions to ask buyers include the following:

- Do you already own alpacas? If yes, what sex and age are your existing alpacas?

- Do you own other livestock? If yes, will this livestock live with the alpacas?

- How much land is available for the alpacas' usage?

- What type of shelter do you have?

- What type of fencing do you have?

These questions will help you decide if the buyers can adequately care for the alpacas they would like to purchase. It's important to ask, listen, and help educate when needed. Many buyers will welcome this opportunity for deeper learning!

I qualify all buyers *hard*. I've had some individuals get annoyed with me, and that's okay. My job is to protect my animals and make sure they are going to a good home that is suitable and capable of providing proper care.

What I've also found is that the individuals who like my questions and respond well to them are also the people who will take good care of my alpacas once they leave our farm. These buyers appreciate my due diligence, and they have said it helps them know they are buying from a reputable farm that truly loves its animals. Those are the people I want to sell to, because I know we are like-minded and they will give my alpacas the love they deserve.

CHAPTER 17

Marketing Your Farm in a Digital World

Every day, each of us will take some sort of customer journey, likely in both the physical and digital worlds. Decades ago, our customer journeys were limited to storefronts and physical goods. Today we live in a multimedia world of websites, social media, chat sessions, and email. These digital platforms have changed the way we interact with businesses, how we evaluate what they offer, and how we make purchases.

When the world moved to digital, we were presented with a massive amount of new challenges but also a plethora of opportunities for farms and small businesses. We are no longer constrained to physical storefronts or in-person visitors. We can now reach anyone, and they can be anywhere in the world. This makes digital marketing a powerful asset that, as an alpaca owner, you can use to reach prospective buyers and sell your animals, products, or services.

To do this well, we need to embrace the good in what we do and what we offer. We need to be open to providing information about ourselves and our farms so we can reach those who are looking for what we have to sell. For some of us, this level of openness is really hard. Inside each and every one of us is a little voice that questions our own greatness. We might have doubts about our farm's ability to achieve success, or we question the strength of our alpaca herd, our products, or our services.

The alpaca industry as a whole also struggles with embracing the latest modern technology, which limits a farm's ability to successfully market and sell online. Our day-to-day priorities are generally focused on the care and well-being of our animals and keeping the farm running smoothly. There are only so many hours in a day, and marketing and branding often end up taking a backseat.

I want to remind you of the importance of marketing and the value you bring to the alpaca community, as well as offer some suggestions for building an online brand that will position you to successfully reach new customers and sell them what you have to offer. And you have *a lot* to offer!

Say Goodbye to Imposter Syndrome

On the internet it is all too easy to compare ourselves to the competition. We look at their websites, their alpacas, and their social media accounts only to compare them to our own. In many cases, we feel inferior, which can paralyze and totally derail us.

Whether you are a young hobby farm couple or a seasoned rancher, it is easy to question your worth when browsing online. Social media only amplifies this because the digital world tends to only show the good (and many times the fake).

On a personal level, this is commonly referred to as impostor syndrome. It has many definitions, but at its core, it is the inability to embrace or accept success. Regardless of what you call it, we all struggle with it at one time or another.

I often see imposter syndrome in the alpaca industry when the subjects of technology and marketing come up. For example, most farms rely heavily on template marketplace websites rather than trying to create their own website.

You might be thinking it's easy for me to make such statements since I have been working in marketing and technology for decades. That's true, but I didn't start there, and I struggled a lot along the way.

As a toddler I stuttered and had an incredible fear of people. In elementary school I was in remedial reading, and I was well behind my peers in all subjects. Growing up, I was a ward of the state and relied on governmental aid. I was the forgotten child and the one who no one expected to succeed. I may appear confident now, but I still question myself just like everyone else does. When I look at other alpaca owners' social media activity, I need to remind myself that I have my own superpowers and to be thankful for the skills I do have as well as the success I have found for our farm and business.

It is easy to forget your superpowers when you are struggling with something outside your comfort zone. For instance, I have to remind myself that it's okay if my blog post has a typo. My online brand is about me, my farm, and my alpacas. It isn't about the typo I didn't see and fix. The truth of the matter is mistakes make us human, and they help people remember there is a real person behind the blog posts, presentations, or product offerings.

The only way I've been able to defeat my impostor syndrome is to focus on what I am good at and how I can turn that skill into something that could help others. As for the skills I'm not good at, I've learned to collaborate with others or hire professionals to finish what I could not. I'm happier for it, and so are my alpacas.

If you are stumbling to launch your website or social media presence, know that you can take the same approach. You just need to find your inner superpower. Figure out what you are good at and use those strengths to help others. Your superpower will overtake the impostor syndrome, and before you realize it, you will be creating something successful.

I've dedicated an entire section of this chapter to imposter syndrome because it is by far one of the biggest roadblocks I see in the alpaca industry. Unfortunately, very few people are knowledgeable in this level of technology, and this lack of experience means we miss out on opportunities for growth.

But it doesn't have to be that way. I want you to try, I expect you to stumble—all of us do—and I know whatever you produce will be better than accepting the status quo.

Everyone Needs a Website

I built my first website in 2000, and let me just admit that it was *absolutely* horrible. I had no prior experience and just fumbled along until I had something that could go live. Once it did, the company I worked for started receiving inquiries, and those inquiries resulted in customers spending money. That little adventure gave me the courage to forge ahead and keep trying. With each iteration I did better, and I learned something every time.

Your farm website doesn't have to be perfect, and it doesn't have to look like a high-price agency built it. What it *does* need to do is tell your story. Your story is powerful, and when you let that story come out, people will connect with you.

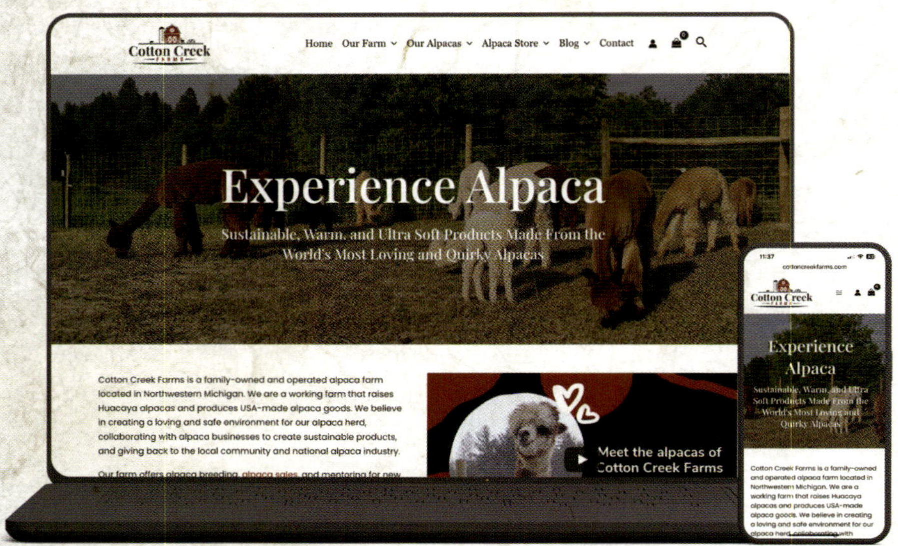

Our Cotton Creek Farms website is built with WordPress, and we use WooCommerce for our online store.

Options for Building Your Website

There are many options for building your own website, and there is no perfect solution that will fit every business owner. Some of the most popular choices for alpaca farms and ranches include:

- **WordPress:** WordPress.org is a free software solution that allows you to create your owner website with pre-built templates called "themes." While WordPress has a lot of base features as part of the core software, it also has about fifty thousand plugins that serve as functional add-ons. If you'd like to use WordPress for your website, I recommend SiteGround as your hosting company. It is inexpensive and very reliable.

- **Squarespace:** Squarespace is another great option for small businesses. It won't have all the options that WordPress offers, but this will make it easier to use for novice website owners. Instead of needing to purchase hosting, you'll pay a monthly fee directly to Squarespace.

- **Wix:** Wix is similar to Squarespace in many ways. You'll be able to make your website your own, but you will be limited in functionality in comparison to WordPress.

- **Openherd:** Openherd is a website platform that is very specific to livestock. It offers some good recordkeeping options for managing your herd and is easy to use. The downside is that every Openherd website tends to look alike, you have very limited functional enhancements available, and Google doesn't embrace Openherd in the same way it does a WordPress, Squarespace, or Wix website.

What to Include on Your Website

Before you stare at a blank web page and feel that imposter syndrome creep in, let's review what you need to cover within those pages. Answering a few questions can help you think through what information you need to provide to your website visitors.

Who are you? This question doesn't need a formal reply, a LinkedIn profile description, or anything fancy. Just think about the words you would use to describe your farm to a friend, neighbor, or visitor. How did you discover alpacas? Why are you passionate about them? What is your background and how does this influence what you do on your farm or ranch? How did that discovery impact your life and influence you to purchase a herd of alpacas?

What is your why? Your why is your purpose, your cause, or your belief in something bigger than yourself. People don't buy your what—they buy your *why*. You own alpacas or you are considering buying alpacas because you've connected to them in a way unlike with any other animal. Share the reasons you offer what you do and the why behind it.

What do you do and offer? What can someone expect from a visit to your farm or ranch? Can they take a tour, visit a store, participate in yoga, or buy an alpaca? Can they board their own alpacas? Can they expect mentoring? Do you handmake your own products for sale? Are they made from your herd? Describe your offerings so that website visitors know what you do, which will help them decide if your business is a fit for them.

How can visitors engage with you? Where are you located and when are you open? How would you like a website visitor to interact with you if they are interested?

Would this be a phone call, an email, or an inquiry via a form? Do you want them to be able to purchase a tour ticket online? Overcommunicate with your digital audience so they can know what to expect and best plan their interactions with your farm.

Answering these questions will help ensure that you provide a lot of information for your website visitors. It will help them get to know you, help you craft your brand story, and help website visitors know if you're a fit for their needs.

Now that you've created some solid text to put on the website, let's walk through some of the additional things you should include on your page. I'll list these items in the traditional flow of website navigation (the menu bar), where you would start with an About page on the far left, then flow into what you offer, and end with contact information on the far right. It's best to keep your contact information on the end so it jumps out at users. If you already have a site, consider this an audit list.

Farm overview with your story and your why. I've found alpaca owners to be incredibly passionate about the animals, the industry, and their business. Let that passion come through so website visitors can sense your enthusiasm and feel encouraged to reach out!

Your differentiators that highlight why you're different from your competitors. Be careful about stating you have championship bloodlines or superior quality. Virtually every show-focused alpaca ranch states this, which makes these kinds of statements generic.

Farm photos to help people feel more connected emotionally. Words are powerful, but a photo has a lot more impact. Use images to tell your story. Photos do not need to be professionally done; in most cases, a smartphone will work perfectly.

A list of alpacas for sale, with as many details and photos as possible. We covered this in the previous chapter, so I won't expound on it here. If you need inspiration, visit a lot of farm websites and ask yourself which alpaca profiles create interest.

A list of services you offer. These services could include breeding, mentoring, boarding, tours, events, photo shoots, alpaca rentals, or birthday parties. Don't just list the item—also provide information on each. You could have a simple paragraph per service, or a page dedicated to each one. My recommendation is to start with a paragraph and work yourself up to a page if warranted.

A list of any scheduled events. Don't just rely on Facebook to market your events. There is a large part of the population who doesn't use Facebook and you don't want to lose out on this demographic. Many families plan travel months in advance, so don't wait to get things scheduled and listed online. I generally make sure we have our summer tour schedule ready for bookings in March. If I don't, I start to get tons of calls and emails asking for that information.

If you offer tours and events, provide visitors with the ability to book online. Some people will plan months in advance, while others will make decisions at night once their kids are snuggled into bed. In both cases, people want self-service, and many want to be able to plan or purchase tickets without having to call. While younger generations are great at using technology, they hate talking on the phone. If you require a phone call to book a tour, you'll lose these generations quickly.

Promote your products and store. If you have an online store, get as many products up on it as possible. If you only have an on-site store, make sure your website offers photos of your physical store so people know what you have and what is waiting for them.

If you offer on-site activities, include easy-to-follow directions. We are in the middle of two cities and close to a popular ski resort. Our directions page includes a Google map as well as directions from the cities and the resort. Before having this information available, we received nonstop calls for directions. Once I added this page, the calls stopped and people were a lot less stressed upon arrival.

Include contact information and offer a variety of methods. Every farm website that welcomes visitors should include a phone number, an email, and possibly an inquiry form.

Let's Talk Social

Social media is a great way to create a *brand*. Branding doesn't need to be elaborate. It just needs to be consistent and authentic. It needs to connect with an audience and tell a story. The brand needs to be recognizable, and it needs to be a good representation of who you are and what you do.

As you begin to use social media to build your brand, remember that artificial intelligence and altered content are everywhere. Influencers have given the impression that we need to live in a world of edited images and perfect lives. The reality of the situation is people don't want perfect. They want *real*. Be genuine and allow that to flow through all you do on social media.

Most social media users want to see everyday farm life.
What may seem mundane to you can be exciting to others.

My most impactful social posts have been ones where I've shared our real farm life experiences. Those moments when nothing goes right, where we fall and have to pick ourselves up to push through the day, such as when a cria got stuck in the birth canal or when our farm favorite, Grandma Amber, died. These stories resonate because they are genuine and people can relate to them. Focus on bringing real into your posts and don't worry about being perfect.

As you consider what social media networks to use, keep in mind there are lots of options and you don't need to use all of them. Start with one and see which platform works for you and your business.

Here are some social media options to consider:

- **Facebook:** Facebook is the third most visited website in the world, but the most popular social network when you look at total activity (both app and website combined). It is a good option, but you need to set it up properly from the beginning. You want to create a *business page*, not a personal profile. Business pages allow people to like or follow you, which skips the need to friend you. Facebook works great for creating events and getting word out via attendees automatically sharing the event for you. Facebook also offers several industry and community groups that you can use to communicate with others who share the same interests. Just be careful not to flood groups with posts or share irrelevant content. This will lead to people blocking you or groups removing you.

- **Instagram:** Instagram is the fourth most visited website in the world, and it has a loyal following with younger generations. It is very visual, and while it used to cater to images, in more recent years Instagram has favored videos.

- **Pinterest:** Pinterest is great for farm stores. While it can be used to promote informational content, I've seen the best success with sharing photos of our products. It isn't hard to do, and many website software tools have add-ons to make sharing to Pinterest super easy.

- **TikTok:** TikTok is an all-video platform that has had massive growth since its US debut. TikTok is great for building a brand, and younger generations prefer it for product reviews and recommendations.

- **YouTube:** YouTube is the second most visited website after Google. It often dominates the Google search results page and has a very loyal user base. Videos used to be long form, but in recent years YouTube has rolled out "shorts" in its effort to compete with TikTok.

I'm going to admit that I dislike Pinterest and I really hate TikTok. I have secured a place for our farm on both, but I don't actively use them. I simply don't like the networks, and I've learned it is really hard to be good at something you hate. Because of this, I tend to focus more on Facebook and Instagram.

Creating Local Search Profiles

Local, map-based searches are a powerful tool for alpaca owners who wish to encourage visits to their farm or ranch. This option only requires you to set up a few online profiles and complete the necessary fields. You won't need any technical skills, which makes this a great option for augmenting your digital marketing efforts.

Local directories are the modern equivalent to the old-school Yellow Pages. They help search engines understand more about your business, and they help humans discover local resources for alpacas, products, and services. The best part is they are all free!

Each profile will ask for similar data that includes:

- Business name, address, and phone number

- Hours of operation

- Basic description

- Website link

- Logo or profile image

Focus on three local profiles:

- **Google Business Profile:** When a user performs a search on Google for locally available products or services, a map with listings often appears at the top of the search results page. These results originate from Google Business Profiles. Setting this profile up will help Google better understand your business and what you offer.

- **Bing Places:** Bing Places drives search results via a map on Bing. You'll claim your listing and complete the fields available within the profile. Once completed, your business will have an opportunity to show up in the "near me" results via Bing's integrated map.

- **Apple Business Connect:** Apple Business Connect is a profile that allows you to appear in results on Apple Maps, Wallet, and Siri via devices like iPhones and iPads.

While there are many other map-based directories available, Google, Bing, and Apple will help you cover the majority of local search activity.

Just Start Creating!

Throughout this chapter I've tried to present options for digital marketing, but also reinforce that perfection isn't required. The beautiful thing about the digital world is that it is easy to change, fix, and augment what you've created. The online world isn't free from imperfections, so do not expect your work to be either. Just start creating *something*. You can always go back and fix what you don't like or improve it as needed.

The important thing to remember is you've created something amazing with your farm and the world wants to know about it. Alpacas are adorable and likable, and they tend to sell themselves.

Open your ranch up to the digital world and you'll be surprised at how many people will be happy with what you've offered.

AFTERWORD

Thank you for letting me share my alpaca journey with you! I am so grateful for the alpaca lifestyle I have found and the gifts it has brought our family. I truly want others to experience that same happiness and peace, which is why this book is being published.

For our family, what began as a purposeful pivot to start a farm turned into an amazing alpaca adventure that is filled with friends, community, and profits. I've found alpaca ownership usually stems from a need that closely aligns with a specific stage in life or a particular set of life goals. The Gill family was no different. We were fleeing the fast-paced life of the suburbs in search of dirt roads and a slower pace of life.

When future alpaca owners visit our farm, they are also searching. They are young hobby farmers on the hunt for kid-friendly livestock, city dwellers looking for nature, or retirees seeking tax deductions or post-retirement business opportunities. The common thread that weaves us all together is the desire for something better. Alpacas bring that and so much more.

I hope my information and personal stories have helped you see the magic alpacas can provide and the amazing adventures they can offer. With each year, my husband and I discover something new and immerse ourselves further into alpaca ownership and the community.

PHOTO CREDIT: DAVE SPECKMAN FROM SPECKMAN PHOTOGRAPHY, LLC

The beautiful thing we've discovered is there isn't just one way to raise alpacas. You get to decide what alpaca lifestyle fits you best. And I promise you: No matter what path you ultimately take, your alpaca herd will reward you with love, entertainment, and a sense of peace you didn't know was possible.

I wish you all the goodness, kisses, and laughter I know alpacas can bring.

SUGGESTED RESOURCES

Associations

Alpaca Association Benelux (AAB)—alpaca-benelux.com

Alpaca Association of Ireland—alpaca.ie

Alpaca Association of New Zealand—alpaca.org.nz

Alpaca Breeding Association Germany e.V.—azvd.de

Alpaca Livestock Producers and Cooperators Association—alpaca.ca

Alpaca Owners Association, Inc.—alpacainfo.com

Australian Alpaca Association—alpaca.asn.au

British Alpaca Society—bas-uk.com

Canadian Llama and Alpaca Association—claacanada.com

Danish Llama and Alpaca Association—dlaf.dk

French Association of Llamas and Alpacas—lamas-alpagas.org

International Alpaca Association—aia.org.pe

Llama & Alpaca Registries Europe—lareu.org

New World Camelids Switzerland—nwks.ch

Norwegian Alpaca Association—alpakkaforeningen.no

Polish Alpaca Breeders Association—pzha.pl

South African Alpaca Breeders' Society—alpacasociety.co.za

Facebook Groups

Alpaca Everything—facebook.com/groups/873886269299399/

Alpaca Farm Life—facebook.com/groups/Alpacafarmlife

Fiber Testing

Alpaca Consulting Services USA—alpacaconsultingusa.com

Société Générale de Surveillance SA (SGS)—sgs.com

Supplies

Jeffers—jefferspet.com

Light Livestock Equipment and Supply—lightlivestockequipment.com

Quality Llama Products—llamaproducts.com

Tractor Supply Company—tractorsupply.com

Training and Behavior

CAMELIDynamics—camelidynamics.com

INDEX